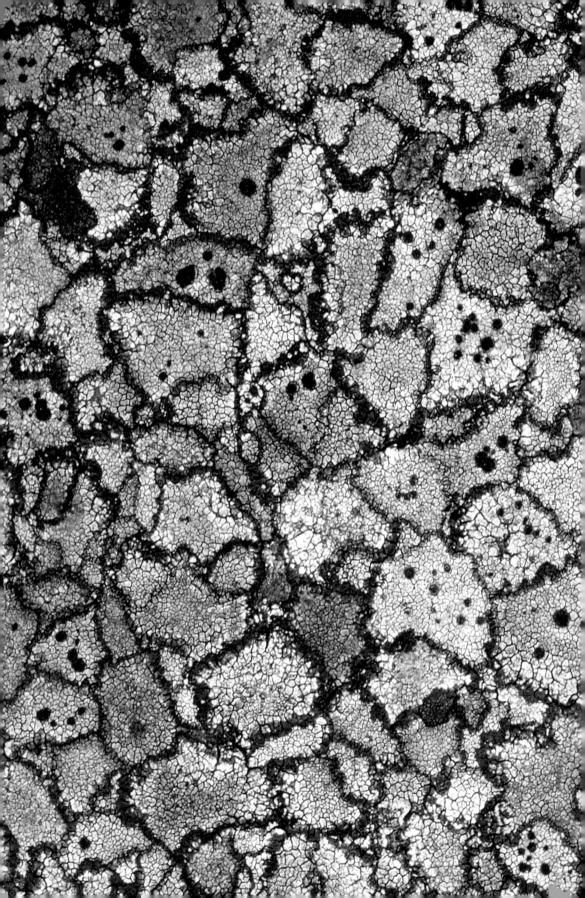

Eagle Days

Eagle Days

Stuart Rae

LANGFORD PRESS 2012

Langford Press, 10 New Road,
Langtoft, Peterborough PE6 9LE

www.langford-press.co.uk

Email sales@langford-press.co.uk

ISBN 978-1-904078-44-9

Design origination and typeset by
MRM Graphics Ltd, Winslow, Bucks
Printed in Spain

Front cover top: A top predator's eyes; an eaglet stares wide with steadily focused eyes.

Front cover bottom: Walking on a high ridge in the Highlands, ever watchful for eagles.

Frontispiece: An eagle looks down from its high regal perch.

Back cover: A golden eagle chick strikes a heraldic pose.

Lichen papers: Whenever I'm in the hills, even if the rain is pouring down my neck, I always see something of interest. It might be a colourful sky, a dramatic landscape, rich woodland, some animal behaviour, or as in this case a mosaic of crustose lichen species. These lichens have gained hold on the smooth surface of a broken piece of Torridonian sandstone. The thalli have spread to form a continuous crust and their prothalli, the dark borders, create a wonderful pattern.
Many lichens are difficult to distinguish so I shall have to go back someday and follow an identification key, examine samples under a microscope or even use chemical spot tests to identify them. A photograph is not enough.

To Di

Who was there on the first day, and the last

Even though eagles are large, scale is difficult to discern when they are high in the sky.
Here a hooded crow shoots away after taking a dive at a golden eagle,
and shows just how big eagles are.

Contents

Foreword   xiii

Introduction   xv

Acknowledgements   xix

1   New Year Days   1

2   Spring Display   9

3   Hunting   17

4   Capercaillie   31

5   White-tailed Sea Eagle   35

6   Still Dawn   47

7   Equinox   53

8   Nest Building   57

9   Incubation   67

10   Mountain Plants   75

11   Change Over   83

12   New Life   89

13   Short Cuts and Otters   95

14   Eagles in the Mist   101

15   Thermal of Birds   113

16   Ringing Chicks   117

17   Prey   129

18   First Flight   141

19   Caledonian Wood   147

20   When the Geese Come In   153

21   November Days   159

22   Snow Storm   179

23   Timelessness   187

Further Reading   193

Winter ridge-walking in perfect conditions on firm squeaky snow, with low cloud below and blue skies above. The sighting of an eagle would be the icing on the cake and truly make the day. One to remember.

Foreword

The golden eagle is a symbol of the Scottish Highlands, an area renowned world-wide for its landscapes and wildlife. To many people, a glimpse of a golden eagle is a once in a lifetime experience. It makes their day. I have watched the joy on companions' faces when seeing such a bird. These people have been tourists, birdwatchers, hillwalkers, climbers, or hunters after red deer and grouse. Whatever their reasons for being in the Highlands, they all welcomed and were thrilled by the sight of an eagle. Such days are special, and those people will remember their eagles forever.

I have seen hundreds of eagles, and watched all aspects of their behaviour. Yet, I can still remember, well if not every one, most of my own personal encounters, as they have all been different. In this book I describe just a selection of those special days in the hills - eagle days.

Stuart Rae

Publishers note:
The views and opinions within this book are those of the author, based on many years of field work.

I always talk softly to eagles, adults and chicks, when I approach them on the nest. That way, they are prepared for an intruder coming up over their nest branch or around the cliff edge. And I continue to talk quietly as I hold them while ringing and measuring them. They don't understand a word I say, but they might be a bit more relaxed, for I have never heard a predator talk to their prey.

Introduction

This book is based on many years of experience with eagles.

My first encounters with them were in my schooldays when I was beginning to explore wildplaces and become conscious of the wealth of wildlife in the hills of north-east Scotland. The taste of those first adventures and discoveries can never be repeated, they were the freshest, and I'll never forget them. I was fourteen when I found my first eagle eyrie and it is undoubtedly on one's first close up meeting with an eagle that their spell is cast. Yet I was inspired by the landscape as much as the eagles, and from then on I have been inexhaustibly lured to travel around the hills and study the wildlife in them.

The concept of this book came from various friends over the years who suggested that I should put my experiences and knowledge down in writing and present my photographs to a wider audience. So the book began to form. It was not written all at once, rather gathered together from pieces written for myself over the years after days in the field, supported with extracts from diary pages and field notes. Some topics are better illustrated with photographs, and I have tried to use those which capture the essence of the Highlands as well as illustrate a point. And I have mixed in further knowledge of eagles to form a more complete description of eagles, their behaviour and ecology.

My time with eagles has been shared with many people, all of whom I thank for their help and memorable days. Those first encounters with eagles were with my school friends Gordon Mitchell and Mike Reid, and my brother Robert, 'Skitts' to the birders who know him. My early eagle days were mostly shared with my brother, and it took years, but between us we explored all possible nest cliffs and trees in order to compile a complete map of the eagles in the eastern Cairngorms. As our range expanded we met other local eagle enthusiasts, notably Adam Watson and Sandy Paine, who both had extensive historical knowledge of the birds' use of the sites we were studying, and more.

Around the same time I met Keith Brockie. Not only is he a highly competent wildlife artist, he is also a very observant and perceptive field ornithologist – more so than some professionals. I have spent many long days in the hills of Angus and Perthshire with Keith, checking high corries for hidden eyries and pioneering routes up or down cliffs to ring chicks.

Our combined knowledge of the eagles and all other raptors in the area grew rapidly during those years and we soon realised that it would be better for the birds if we coordinated our efforts. By doing so, we could ensure that the birds were visited only as often as necessary each year, we could pool our results and all the sites could be surveyed efficiently each year. Consequently, together with a few other raptor enthusiasts, the North East Raptor Study Group was formed, the first to be set up in the country.

Then in the early 1980s I was offered the opportunity to study golden eagles with Jeff Watson and Derek Langslow of the Nature Conservancy (now Scottish Natural Heritage). I took it. And I am ever grateful to them for some of the best days of my life. Not only had I the best job, but the best boss too. Sadly, Jeff died in 2007. He was always humble in expressing his own efforts, while most appreciative of my own; paying heed to my ideas, and trusting that I would collect the necessary data which required walking hundreds of miles in the hills in all weather.

We mostly worked separately, alone in remote places, so during those years I was thankful for many a night's shelter and kitchen table banter with Davie and Isabel Duncan. Those conversations included lots of snippets of local information on various issues of Highland wildlife, and over the years I have gathered a trove of such miscellany. In the interim years I have explored most of the Highlands and Islands, and in my travels I have learned where most of the eagle home ranges and nest sites are, what the birds in different areas eat, how well they breed, and perhaps most important of all in the human perspective, how eagles fit into the landscape and the associated land-use.

Land use in the Highlands is constantly changing and at the turn of the century I began studying eagles in relation to proposed and operating wind farms; assessing the likely impact of any development on eagles. And other recent work has been with Mike McGrady, whom I thank for teaching me how to fit radio-tags on eagle chicks for satellite–tracking. Once tagged, the birds' movements can be followed on a computer screen back at the desk – a technique which wasn't even considered when I first started my studies. The technology available today is remarkable, and what will be available tomorrow should help advance our knowledge of eagles even further. Yet there is no substitute for good solid ground work to record what the birds are doing when and where. Ewan Weston has recently been helping with the monitoring of some of the north-east eagles, and it is good that we can pass on our knowledge to someone of the next generation. He has since embarked on his own study of eagle movements using satellite-tracked radio tags. And he is also a great help with the climbing.

Currently I am once again working with my friends and colleagues Skitts, Adam and Keith, analysing our long-term set of data on the eagles in the eastern Highlands. Almost all this information has been collected in our own time, thousands of man hours between us, and its value is now apparent. The long-term data are relative to contemporary differences in such factors as food supply or weather and as the information cannot be collected retrospectively, it could be considered priceless. All those days on the hill have therefore been worth it, even the quiet or mundane ones, if they have helped our understanding of the behaviour and ecology of the golden eagle. And above all, have helped conserve a healthy Scottish population.

A few points to ponder before reading this book

No place names are given in this text or photographs used if they could lead a reader to an eagle nest site. When readers go to the hills they are encouraged to watch eagles, or any other wildlife for that matter, from a distance and not to approach nests, nor to tell others if they do find an eyrie. Too often I have heard excited birdwatchers talking to one another in raised voices how they had sat by an eyrie watching eagles and giving the details how to get there to any and everyone within earshot. The eagles' habitat is shrinking all the time; they need space and peace to survive. Eagles and the Highlands are wonderful, let's keep them that way.

Nor are the locations given for all the photographs or scenes described in the text, there is no need. They have been selected for generic representation. If anyone recognises a scene they can reflect on it and empathise with our passion for and knowledge of the Highlands. Those who are not acquainted with any scene can look forward to discovering it for themselves – too much sense of discovery has already been robbed from people's personal adventures by guidebooks and maps.

To go out looking specifically for eagles can be very disappointing as there are many blank days. Far better to go out, enjoy and appreciate whatever one comes across, whether it is the landscape, plants, insects, eagles or even the weather, all are interesting. Although I work mostly alone, I get pleasure from company, and companions have often asked me what something is or why it is there. Sometimes I have not known the answers even though I might have walked past the subject numerous times, and in response I have looked it up when we returned. Nobody knows everything and there is a wealth of fascinating things to learn about the Highlands' wildlife.

Eagles are only one thread of the Highland tapestry, a golden thread.

Acknowledgements

Those central to my life with eagles have been mentioned above and I am forever grateful to them all for their company, discussions and our shared experiences with eagles. However, it is not only eagle enthusiasts or ornithologists who are enthralled by eagles, many lay people are too and I have tried to pitch a text to suit all. So during the development of the book I bounced ideas around a wider group of friends amongst whom Lesley Thorpe and Jenny Mason gave helpful comments on drafts of the book as it developed. As the book neared completion my wife Di and our boys Lachlan and Duncan all read it through and gave useful feedback, expressing their likes and dislikes. Frank and Rosemary Kovachevich honed it further with astute edits and raised good questions which needed to be more thoroughly addressed. Adam Watson then gave some very helpful comments which improved the final consolidation. And I thank Keith Brockie for giving me the opportunity to take the close up shots of a golden eagle used in this book, which are of an injured bird he was nursing.

I shall always remember that day when Ian Langford called me on my mobile phone to discuss the publication of the book. He rang when I was far out on a Highland hillside surveying, among other species, eagles. I sat down in the heather to expand on the concept and he confirmed there and then that he would be delighted to publish the book. So I thank Ian for that special eagle day and his positive attitude towards publishing the book. Then as production progressed, I thank Stuart Fortune at MRM Graphics for his editorial work on pulling the book all together.

Those people who know me will understand how much thanks I owe to my wife Di for tolerating me on all those long days on the hill, and more so for all those long days cooped up in a house with me during spells of wild wet weather. And the boys have grown up expecting and understanding that their dad is likely to disappear into the hills for months at a time. Thankfully my family know only too well how much eagles and what I share with them mean to me - the freedom of the hills.

Opposite:
The eagle is a recognised symbol of strength and power. Its image has been widely used since pre-historical times in carvings, heroic tales, and heraldry. Today it is still used for these as well as in popular art, product labels, company logos and national emblems. This carving of an eagle on a Pictish symbol stone in Strathpeffer is distinguishable as a sea eagle as its lower tarsi are bare while those of a golden eagle are feathered down to the toes. Sea eagles were probably more familiar to early man in Scotland as they would have been more frequent than golden eagles along the lower stretches of rivers and coasts where most of the early settlements were located.

The distinctive silhouette of a golden eagle crossing the sky
Man has looked up many a time
And thought......

1

NEW YEAR DAYS

Eagles are an integral part of the Highlands

Hard blasts of snow streaked across the black basalt cliff and the ground turned white. The Old Man leant into the wind as it scoured The Storr on Skye, and as the Atlantic westerlies poured great clouds over the escarpment above; the cluster of pinnacles curled eddies and drifts in their lee. A pair of ravens flushed with a disgruntled croak from a cosy niche high up on the Old Man himself. Within seconds they were gone, swerving and dipping along with the flow. Clasped in the hand of winter, in all its rage, there was a raw beauty in the dark rock textures high-lit by snow, and in the way the Old Man stood stark naked with the rest of the pinnacles, all before the vast complex of pillars and gullies of the cliffs, the part-etched pinnacles of tomorrow.

I spent a wonderful morning exploring the various nooks and crannies and all their aspects. But it was cold, finger-nipping cold. I had to keep moving, so I ran back down to the car and drove on north, to the Quiraing, where the cliffs are a touch smaller and the wind was less fierce. A short steep walk and a scramble led me through a line of narrow rocky passages to the inner sanctuary, where I escaped from the wind completely, and I was surrounded by a silence, that unique mountain silence.

The Table, a natural level platform of about seventy metres width lies at the heart of the Quiraing, and that day it was spread with a pristine cloth of snow.

The Old Man of Storr

- one of a group of spectacular basalt pinnacles that have been left by landslips and erosion below the cliffs on the Trotternish peninsula of Skye. Such escarpments create fine steady updrafts for eagles to soar on.

Looking out through the surrounding ring of pinnacles, a yellow light cast across the various knolls on the moor below. The grass shone bright bronze and twinkled as it ruffled and waved. Out on the Minch, white water chopped the surface and no boats had defied the gale warning. Staffin Bay at high tide showed a glimpse of its bright sands set in a crescent, the curve emphasised by the radiating dykes and fences of the beachhead strips of croftland. At the head of each strip lay a white-washed cottage with its back to the moor and the wind. Not a soul was about.

Far over on the mainland, beyond the Minch, the snow was down to about one hundred metres, separating the high cliffs of Torridon from their pedestal of pale mauve moorland. Baosbheinn, in particular stood out clearly, its long ridge stretched away north from the tight group of Beinn Alligin, Beinn Eighe and Liathach. Loch Gairloch showed up blue in the returning light after the worst of the black clouds had rolled on their way. Loch Torridon to the south of the hills was receiving them now, sitting in gloom under thick falling snow.

After poking around and investigating every angle of every buttress and gully, and pondering over the drops, I left the amphitheatre via the red rocky shoot that twists around the foot of The Needle, the most spectacular pinnacle. The rock rises so steeply that it overhangs in places on each side. The rock is fragile, often dishonestly so and the top is particularly thin and precarious, presenting a point barely wide enough for the sole of a boot. Yet, it has been climbed. I looked at and thought about it. I even traced out a possible line, but that was frightening enough so I went on my way.

Back in the wind, I stopped for a while in a hollow beside the low-lying wall of rock known as The Prison. I spied along the line of cliffs on the southern horizon with the binoculars and picked out the distinctive profile of a golden eagle sitting on the very point of a hill. In the low January backlight it cast a sharp silhouette of broad shoulders, heraldic head and a deep hooked bill. It was faced into the wind, twitching its head this way and that, watching all around. I blinked; it spread out its wings and was gone. The bird had dropped down the far side of the hilltop, caught the wind under its wings and lifted high in the air. Turning and gliding it was soon fifty metres higher than its perch without a single flap of its wings. The massive wings of the golden eagle provide a fine lift when a good breeze blows across their two metre span and half metre depth. A flick of the tail or a twist of a wing tip is all that is required to control the bird's flight.

The wings closed tight to the bird's body; and dipping its head, the eagle plunged into a great long dive. Ripping down through the air it pulled open its wings and almost as quickly as it descended, it rose on the wind and regained its height. Just as it teetered to a halt it repeated the dive and the bounce. And then it did it once more. Having lost a little height on each dive and lift, especially on the last one,

The Table in the Quirang on Skye

- a popular tourist and hillwalking venue.
Eagles also like to visit there, and hunt the rabbits that live on the lush grassland growing
on the rich basalt-based soil.

the eagle curled in a wide sequence of loops to return to its perch. Throughout the whole performance the bird had not beaten its wings.

That had been a dramatic deep diving display which was performed for another eagle, either its mate or its neighbour which would have been watching the show from somewhere unseen by myself. Although truly dramatic to me, it was all in a day's work for an eagle.

The Quiraing with all its spires and towers is perhaps the weirdest of warrens. The basalt bedrock provides rich nutrients for the grasses. That is why the grass there is so lush - and why it is so heavily grazed. However, although the grazing is plentiful in summer, in winter there is less growth and corpses of sheep and rabbits are common in these hills. So, there is a good supply of live rabbits in the summer and sheep and rabbit carrion in winter. This provides food for the eagles, buzzards and ravens all year round and all are ever present.

On another day in the Quiraing when an eagle was up, a thin mist skimmed lightly across the plateau and the pinnacles scraped their points on the ceiling. There was no snow lying and the table was a rich billiard green. Ever nibbling sheep and rabbits crop the grasses short, making it treacherously slippery when wet, so I had to watch my step and keep back from the edges. Instead of going into the Quiraing, I traced the rim of the escarpment around and above. A pair of shy buzzards backed off as I passed their stretch of cliff, mewing in alarm or complaint, but keeping their distance. Unlike the ravens. They cronked and prukked all around me, bobbing and dodging from this side to that as they followed and harangued me until I was well out of their territory. Then they slipped back to watch from a bluff.

With the table set below me, and now that the mist had lifted for a while, I sat and viewed the wide array of far-flung hills. To the south, the rugged Cuillin were in the all too frequent cloud that deprives so many visitors of their truly spectacular skyline. Across the water, the Torridon hills stood round and red faced. Farther north, the Sutherland hills and farther south, the Knoydart hills all faded into the soft blue mist of infinity.

A raven prukked in excitement, but not at me. There was a pair of them again, flying hard and fast down over the edge of the cliffs. I stood up and looked over to follow their course, when around swung an eagle, a fine adult, barely thirty meters away. It seemed to be trying to sneak away from the raven's attentions, softly flapping its wings as it slipped through the pinnacles. There were no young in the raven's nest, the season was wrong. Yet, they scolded and chivvied the passing eagle none the less. The eagle was well around the corner before it had pulled away from the ravens and adopted a more casual flight. Or was that simply an illusion caused by the speed at which an eagle can fly? Eagles are seldom seen close enough to appreciate their true speed. Usually they are seen as mere dots on the horizon or up on high, with no accurate scale.

Ravens habitually harass eagles wherever they meet. They are one of the few bird species that live in the open moorlands throughout the year, and to some extent they compete with eagles for food and nest sites. They are great tormentors of their larger, more dignified adversaries.

The most intense such mobbing that I have seen took place in the Outer Hebrides. I was sitting on a cliff-top above a sea loch where both eagles and ravens nest. Calm evening light shone onto the cliff on an island opposite where the ravens had their nest. It was a restful scene and the ravens had not long flown in to roost. Then suddenly all was in turmoil. There was a clamour of croaks, the echoes of the calls ripped across the loch, bouncing from one cliff to the other and a second pair of ravens turned out from below me. All four birds darted along to the southern end of the island four hundred metres away, sweeping their wings hard and fast.

I had not seen the eagle. It had been sitting quietly on a heathery rise at the farthest tip of the island, less than fifty metres from the shoreline. This is not the most obvious place to look for golden eagles and the bird's dull brown plumage merged into the hillside so well that only its movement gave it away. It ducked its head sharply to avoid the first stoop from the ravens.

He was a blond-headed adult male, and his deep eyebrows appeared to express disgust at the raucous attentions of the ravens. I assumed that the eagle had only recently settled or else the ravens would not have taken offence at his presence so suddenly. Where he had come from I do not know as I had previously been focusing on the ravens through the limited view of my telescope. It had not made a kill there as it didn't try to protect any prey from the ravens. Although, on spying at it through the telescope, I could see that its crop was very full. Perhaps it had gorged itself somewhere on the far side of the loch and glided too low in the slight air, forcing a landing on the island. Such a heavily gorged eagle would find it difficult to gain height, even by flapping its wings deeply.

If a peaceful after-dinner roost was what the eagle was seeking it was not to find it there. All four ravens took turns to dive at its head. They strutted around it, they jumped at it, and they threatened to peck it. One bird tugged out a sod, flew up and dropped it, more in gesture than on target. Several times the eagle was riled enough to lean over on half open wings to thrash out at its tormentors with its wide flashing talons. Those talons stood out as massive and lethal in comparison with the dwarfed blunt feet of the ravens. Yet the ravens persisted in their bold cheekiness. For a second, they tipped the eagle off balance and it hopped downhill to square up. But level ground is not the eagle's domain. It was disadvantaged and outnumbered. It opened up its great wings, ran a few steps downhill and jumped up into the air and half glided, half flapped over the water to land on the hillside on the far side of the loch. It had to fly hard to gain that short distance. If it could have, surely it would have soared up far from the ravens, but it didn't. And the harassment continued till dusk drew the ravens back to their roosts leaving the faint grey shape of the eagle alone on the greyness of the night hill.

I have found remains of young ravens at eagle eyries on several occasions and even those of two full adult birds. So an eagle can certainly kill a raven. However, they are intelligent birds ravens, and those that day probably knew that their lighter agile bodies and wings could out-manoeuvre the eagle while it was on the ground. They knew they were safe from its talons that time.

Back at the Quiraing, I set off down with one eye on the cliffs at my feet and the other on the eagle that had been displaying earlier. It had returned to its perch on the skyline. The dull winter daylight was fading into the darkest of winter nights and the footwork was not easy. While my head was down guiding my feet the bird slipped away and an eagle day was over.

On clear days eagles will soar high, but do neighbouring birds display to one another by soaring, or is that merely a human interpretation?

Eagles are not easily seen when up high, and people seldom look up. They are more frequently and easily seen on clear, windy days when they are more active, flying just above the summit line. On days like that they dive in great long stoops and climb back up again and again on brisk lifting breezes. They are definitely displaying then.

2

SPRING DISPLAY

When people hear that I study eagles, they often think that I have large close up views
of them all the time, like the frame-filling images they see on television. The truth
however, is that I mostly watch mere specks in the sky. There might not be the
same detail seen of the birds themselves, but it is by far the best way to watch
their true, undisturbed and fascinating behaviour.

So many days on the hills begin with a long walk-in before coming upon
anything of interest. Miles of walking are involved just to collect one piece of
information about the eagles. Fortunately, every little piece of information is
always a valuable addition to understanding the birds, their behaviour and their
ecology. And every day is a new day. I never know what I'm going to come
across on my travels. Differences in times of day or year, or weather, all
influence what can happen. Without being there, just being out and about, we
can't expect to experience anything. The only way ahead is to get on and do it.
No matter the time, date, weather or distance.

Never the less, a first venture onto new ground is always more enticing
than covering familiar terrain. Every corner hides something new, perhaps not
new to others, but new to me. And I can particularly remember my first exploration of one Sutherland glen. From the mouth of the glen I could see right along
to the high ground at the top end. There was no need to walk any farther to find
the eagles. They were there, in front of me. They were small because they were

five kilometres away. But, they were perfectly visible. All I needed to do was look and watch.

There were five eagles in view all at once. Four were slowly soaring in spirals, together, in two distinct pairs, and a fifth was flying in towards the group from over the ridge to the east. That bird was most probably a male as it was noticeably one of the smaller birds in the group and it was displaying. As he approached the group, he slowly bounced along in an undulating flight above the dividing ridge. Not in a haphazard sort of flight, but in a recognisably distinct style, which is a form of display. They all rose up together to about a thousand metres, a few hundred metres above the hills. Two of the birds began to dive in spectacular deep dipping displays. They seemed to be doing so in turns. As if one bird was showing how well it could fly, then watched to see what the other could do. Again, by size, it was probably the two males of the pairs that joined in the displaying.

The birds were all in a tight group above the junction at the head of the glen where a pass leads through to another glen and a third glen lies over the ridge on the west. They hung around for about twenty minutes, twisting and jostling, and as such it was difficult to keep tracks of which bird was where and doing what? After the show, after they had seemed to have established who was who and I was totally confused, they split away in different directions. The single bird flew back to whence he came, a corrie in the east, which I discovered later to be the site of its nest cliff. One pair drifted off over to the glen in the west and began to hunt. They dropped in height to about a hundred or less feet above the hillside. And after a zigzagging sequence of swoops, they swung on the wind over the bottom of the steep ground and across the flat glaciated bottom of the glen. It was subtly surprising how much ground they covered in only a few minutes, and how they had gone so quickly from high in the sky to so low in the valley. The second pair floated away over the pass and the sky was once again empty. The air space, where they had all been such a short time before, was probably where their three home ranges converged. The birds had shown each other their flying prowess and sized each other up, but they probably all knew and recognised each other already as eagles are long-lived birds and the turnover of occupants of home ranges is usually very low. There was no contact and no violence. There were enough food and nest sites for them all.

Two weeks later, when I was next in the same glen, there was warm light high above me, and although it was early spring, there was a cold northerly wind and the snowline was down to two hundred metres. I slowly traced the path along under the bastion crags, watching all around me as I went. No sooner had I set off than a raven jumped out of its nest corner low under a dark overhang, calling in alarm at my intrusion. Then a peregrine came off the top of the cliff, screaming in a more raucous alarm. I interpreted that the peregrine's

Hilltops stand above the valley cloud of an anticyclonic temperature inversion. I have often watched soaring eagles from the valley floor, as they lifted higher and higher until they disappeared from my view into and above the clouds. This is the type of view the eagles have once   clear of the top side of low cloud.

alarm was more at the raven than me as it hadn't bothered to call until the raven passed by its nest ledge. These birds are always argumentative neighbours. However, it was good to see them and to know that they were both nesting on the cliff, but my main interest was finding out where the eagles were nesting. I walked on by.

At the far end of the crag, just as the peregrine and raven were calming down, a pair of kestrels added their screeching call to the commotion, and all four birds screamed their heads off. I sped up and after putting a quick kilometre or so between us I was glad to be rid of the din. I crossed a set of sheep enclosures and made a stop to lie down in against a huge boulder below a second and much higher cliff to spy the top end of the glen. Although I had been there so recently before the landscape was still just as stunning. The valley floor was flat and green, and intensified the vertical and clean rocky cliff that reared up ahead. The snow topped ridges beyond were all glistening in the backlight. And there was a fresh smell of spring in the air.

Tucked into the overhangs of the cliff above , there were second pairs of ravens and kestrels. They were cronking and chittering, this time definitely at one another rather than at me as they were at it before I arrived. Each regarded the other as potential predators of their eggs or young. Giant jagged boulders had long ago tumbled to the foot of the cliff and in amongst them were several flat spreads of deer hair and sheep wool. Some animals had sheltered from the winter storms behind the rocks, but never awakened. Now their bones lay scattered by foxes, ravens, crows and eagles, while their pelage felted the ground. Fresh spikes of bright green grass poked through the carpets, enriched by the recycled nutrients.

A male eagle came across the glen from my left and hunted for about half an hour across the opposite hillside. After the quick burst of action, he landed briefly on a hillock, then rose up and began to display in a deep dipping sequence. He glided down in a well-mastered series of curls, opened his tail, lifted his wings up in a tight v and landed in a perfectly stalled motion on a slab of moss covered rock right on the skyline. I dashed quick glances to either side just in time to catch sight of a female as she appeared from the same direction the male had taken. In a very deliberate flight, she gently settled down beside him and together the pair were silhouetted in a blue hazy light. I quickly grabbed and set up my telescope and as soon as I had done so a third eagle, a male, came hunting along the back side of the ridge. SLAM. A peregrine had taken a stoop at him and in a flash of wings, both were quickly out of my sight down the far side. And that was the last I saw of that eagle that day.

After that, things went quiet for a while and I settled into a soft hummock for a long comfortable watch through the telescope. It is in such couched splendour that I probably get close to a television view of eagles, only better, for

I had true wrap-around sound and the fresh sniff of new growth. The reminiscent power of particular scents is something that television will never be able to invoke. Meanwhile, the pair stayed quiet for several minutes before starting a bit of head bobbing and fidgeting.

The male shuffled across to the female and she crouched with her head well forward. He gave a few more bobs of his head and held tight eye contact with her as they stood close side by side. Then he deftly jumped up and landed on her back with his talons tightly closed into fists so as not to harm her. A bit of rearrangement of tails to either side and they comfortably mated while poised on the rock. Their coupling lasted for a minute or so, and all the while, the female kept steady and the male only corrected his balance once or twice with half flicks of his wings. He delicately jumped off and both birds gave head to tail shivers. After that there was much preening of feathers and wing stretching, getting all those important flight feathers back into place. Once everything was returned to shape, the male took a turn, stepped onto the air on the downside of the rock and was off low across the hillside. He was back hunting again.

She took off a quarter an hour later and headed straight in the direction from whence she had come. SLAM again. This time a peregrine, probably the same bird, had swooped and knocked a few feathers out of her back. She tilted down and flew low around the big craggy hill end where the peregrine was obviously nesting and clearly did not want any eagle near.

That turned out to be a very successful day. I went on to find two built up eyries, enough to establish what eagles were nesting where and added the bonus knowledge of two peregrine sites, two kestrel sites and two raven sites. And I had seen several types of display behaviour of four species.

The displays by eagles most commonly witnessed by us are those that involve deep dramatic dips and dives, the undulating display, usually repeated several times and often over twenty times. The most continuously repeated dives I have seen were sixty. Another version of this display is the shallower dipping which is probably performed as another sort of advertisement of home range occupancy. It is performed in one long undulating flight, and might be overlooked as a display if it is not developed into the full dipping form. The revealing feature is the repeating pattern. A third type of display involves birds playing with sticks, stones, grassy sods, or prey items. It is similar to the deep diving form, and like that, can be performed high in the sky, but is more often played out quite close to the ground. One bird, usually the male, will fly up with a stick, for example, drop it, and stoop with closed wings to overtake the stick before it reaches the ground, catch it and re-climb. The bird might repeat this several times, and sometimes the female will do it too. Despite their large size, eagles can be very swift and agile and this display is an impressive piece of flying.

An even more dramatic form of display is talon grappling. I have only witnessed golden eagles grappling once. And that was more than twenty years after I first began going to the hills and watching eagles. Most talon grappling occurs between two male birds in dispute. The birds that I saw doing so were definitely a male and female. I first saw the female soaring over a historical, but recently unused, nest site. Then a second bird joined her. He was a very small male, the smallest I have ever seen. Both birds caught onto a thermal and soared up to about one and a half thousand metres, really high above the highest local hill which stood at six hundred metres. Once up there, they began to display in deep dives. He dived six times, she eight times. They soared around each other for a few minutes as if sizing one another up before he suddenly started diving again, twenty-one times in quick succession. Astonishingly long fast dives, one following onto another. He could certainly fly!

It wasn't easy keeping tabs and it became hectic when a tiny kestrel interrupted the pair. It flew straight in from I don't know where and stooped several times at the female and I was wondering why she tolerated such annoying behaviour when she calmly turned on her back, showed a quick yellow flash of her talons and gave a sudden jink and lurch towards it. With that, the kestrel was off.

All this was happening high in the sky, and if I hadn't picked up the birds first when they were relatively low in the sky, I would never have seen all the action. Once the pair were back soaring together, the male swooped at her. She turned just as she had done towards the kestrel, except this time her talons were met by his. They linked together. And off they went. Down and down, in what seemed to me to be a frightening spin, twisting and tumbling for hundreds of metres. The fall lasted for less than a minute, a very long minute. They tumbled over one another right down to almost ground level, and as the hillside came into the bottom of my binocular view I felt myself hold my breath, and zip, the two birds parted with a nonchalant air and swung in perfectly controlled flights behind the hillside where the nest site was. That was the most amazing display by golden eagles that I have ever witnessed, and it really is amazing what eagles can do. From subsequent notes on the pair's nesting, it would appear that the grappling had been a part of a pair-bonding display. As eagles are long-lived birds, pair bonding is quite infrequent and rarely seen by eagle watchers. All this display happened above the cliff where the two birds settled as a pair and built their first nest. There is no question as to their identities as the female was very big, even for a female, and she was well marked, with white wing patches and tail patch, tawny epaulettes and a rich golden crown. He was very small, blonde, with no white on either wings or tail. And both birds were subsequently seen at the nest. Although most talon grappling witnessed by man has been in territorial disputes, that might simply be because this is the display that is seen most. Perhaps, if more initial pair bonding were witnessed there would be more records of grappling between sexes.

All these types of displaying behaviour can be seen anywhere where eagles live, and that is in most of the Highlands and Islands. I have watched a male diving above one of the most remote glens deep in wild country. He was high above his eyrie with the female watching from the nest where she was sitting on her eggs, with a crook in her neck. And I have watched them above the roadside while sitting in the car, with a crook in my neck. Most eagle activity occurs up in the sky, and the sky is a part of the Highland landscape that is well worth watching.

The most unusual display I have witnessed was by a bird to a helicopter. A coastguard rescue Sikorsky S-92 helicopter was crossing the Isle of Lewis, flying over the moorland at about a hundred metres above ground, when a golden eagle flew out from a nest cliff towards it. The eagle intercepted the helicopter's flight easily and performed six deep dipping displays from approximately a hundred metres above the helicopter, slightly ahead of it and to the nest side. From my notes, the eagle left the cliff at 1558hrs, tracked the helicopter from 1602 - 1603hrs, and then returned to the nest cliff at 1604hrs. All very brief, but intense.

When golden eagles hunt they often ridge-hop; flitting low from one side of a ridge to the other and back again
- a cunning strategy which gives them a chance to catch mountain hares or grouse unawares as they pass over.

3

HUNTING

Man has marvelled at the prowess of birds of prey for thousands of years, more. The hunting skill of these birds is so bewildering, and that of the eagle, such a powerful bird, is truly an amazing sight to watch.

I always keep a loose eye on the skyline when driving through the Highland glens. Occasionally I do see eagles; usually I don't, although I persist because I once had a great view of a bird hunting above a road. I was up in the far north-west when I saw one over a high snow-covered ridge and pulled in to watch it. When I first looked through the telescope I only caught the end of a stoop and I thought I had missed the show. However, I was wrong and I spent almost an hour watching that eagle trying to make a kill.

The eagle had missed on its first strike. So, immediately, the potential prey had gone to ground under a boulder. I didn't see what the prey was, but it couldn't have been far underneath as the eagle kept flying low past the hollow and seemed to be squinting into the shadow. Determined to catch the animal, the eagle landed at the entrance to the cavity and, while leaning back on one leg and flapping its wings for support, it thrust the other leg wholly under the boulder. It stretched and re-stretched, but it couldn't reach its prey. Then it stopped, stood by the boulder, cocked its head and peered into the darkness, and again thrust in a talon. Defeated, and probably frustrated, it seemed to give up. Slyly, the bird flew over to sit atop a snow-topped boulder about fifty metres away and slightly uphill. For the next half hour that bird sat and stared at the black hole in the snow, waiting, just waiting, for whatever was hiding there to chance its luck and make a break from its besieged position. Then with a lunge, the eagle gave   another quick dive at the hole. It must have seen a movement within reach of the entrance, but again it missed. It gave another brief struggle and stretch, and then retired to sit out the siege. I left it there. I had other things to do and had my lunch packed in my rucksack.

I never found out what the prey was, probably a ptarmigan as they will scurry into cover after being chased by an eagle. Or it could have been a mountain hare; these are the two animals most likely to have been up on a mountain top in winter.

The determination of that eagle emphasised the relative scarcity prey in general in the north and west Highlands compared with their abundance in the east. I have often watched eagles in the east miss a grouse or a hare. They miss more often than kill. Rather than persist in an energy-consuming chase, which they would have little chance of winning, the eagles there simply regained their height in the air and continued to quarter the hillside in search of another grouse or hare to surprise. Surprise attack, either by dropping from the air, or occasionally from a perch, gives the eagle its best chance of catching its prey.

When eagles are hunting they glide slowly back and forth over a hill for hours before they even made a dive at anything. To see an eagle make a kill, it is necessary to be very patient and to watch the bird constantly. Or to be lucky and witness a successful strike soon after first seeing the bird. In all the hundreds of days that I've spent with eagles, I've only seen a handful of successful kills. Back in the eastern hills, one such kill was by a bird which plunged into long heather and killed a red grouse. The eagle rose with the grouse immediately and flew over the rise to where I knew there was an eyrie containing young. Within minutes that bird was back and had rejoined its mate, which was still hunting the same area of moorland. A second kill was procured in about a quarter of an hour and that too was dutifully carried off to the eyrie. That is not to say that all hunting is so easy for eagles in the east. I have watched other birds there, even the same individuals as those described above, hunting for hours and not witnessed a successful strike.

In my student days I used to take holiday work as a ghillie in the deer stalking and I often saw eagles while out on the hill all day. One day we had an exceptional view of one. Bert Hardie was the stalker I was with, but I cannot remember who the guest was that day. We had stopped for lunch on a high ridge looking out over an expanse of heather moor, when whooosh. A red grouse shot over our heads, loud and low. Then it turned on a sixpence and dived into the heather just a few metres in front of us. We sat there pop-eyed, still wondering what that was all about, when whoosh again. And an eagle swung

The golden in the eagle's name refers to the rich colours on its head and nape.
Heavy eyebrows shade the eagle's eyes from the sun. This must be a great advantage for
watching prey from high in the sky. They probably also give valuable shelter to the eyes from
stinging rain, hail and snow when flying in the Scottish skies.

over our heads, a little higher than the grouse but still close. It turned and waited on, right in front of us, until gradually it drifted down and along the hillside, all the while scrutinising the ground for any sign of the grouse it had almost caught. Its eyes were so intense, totally focused on the job in hand. I could appreciate the grouse's fear.

The grouse had been chased over the ridge by the eagle and when it saw us it seems to have quickly assessed that we were of less risk than the eagle, so had deliberately taken shelter close to us, also probably correctly assuming that the eagle would not dive after it so close to humans. I walked over the ground where the grouse disappeared, but could not find it, nor did it flush. It had either clapped tight to the ground and daren't move or had crept quietly away under the heather.

An eagle's hunting success depends on several variables; the weather, time of day, time of year, the bird's experience, and most importantly, the availability of potential prey in the bird's hunting area. When all these variables are at their least favourable, during cold wet weather, in a short winter's day, a young inexperienced bird has little chance of catching any live prey, and adult birds do not have much better odds. Fortunately for the eagles, during these conditions they can endure long fasts of several days. Meanwhile the same inclement weather kills the weak or old deer and sheep that have little shelter from the elements out on the open hillsides. This is especially helpful in the west where there is often an abundance of carrion by the end of winter.

This larder is fully exploited, not only by the eagles, but also by buzzards, ravens, hooded crows, foxes and badgers. I have seen three sheep carcasses of recent origin, lying side by side on a hill. Only one had been opened and eaten, right down to the bones and fleece. Of the other two, only the eyes, nose and ears were missing, the work of the crows and foxes. The opened carcass showed evidence of having been eaten by eagle, fox and crows. The eagle had left great white splashes of droppings cast around the heather. The bones had been chewed down by the fox and I flushed a group of ravens and hoodies from the scene as I approached. Foxes usually feed at night, so probably avoided competition with the eagle which feed by day. And the ravens and hoodies would only have been able to feed on the flesh after the others had opened up the skin, as they are not strong enough to open that themselves. Most of the flesh had been eaten and I expected that another of the carcasses would have been opened up fairly soon after I left.

Feeding on carrion might be a safeguard for times when live prey cannot be caught, but as a main diet for the whole year, it is deficient in nutrients and undependable in amount. Far fewer animals die during the summer months, and the flesh of those that do quickly rots or is eaten by other carrion feeders. Those mentioned above, and those that appear only in summer: beetle larvae and the blow fly maggots.

As hares and grouse are scarce in the west compared with those in the east, the eagles in the west take a wider variety of species to supplement their diet. Rabbits are locally abundant and usually accessible to eagles as they can be found out on the hill as opposed to being usually confined to lower ground in the east, where eagles very seldom hunt. In the days pre-myxomatosis, rabbits are said to have been the staple diet of golden eagles in some parts of Scotland, especially in the west. And the decline of the eagles' breeding success in these areas where they used to feed on rabbits might be attributable to the decline of the rabbit population. Certainly, where there are pockets of high rabbit populations, the eagles tend to breed well. Some areas are now experiencing an increase in the numbers of rabbits and the eagles appear to select them as ideal prey. Studies in other countries support this preference of golden eagles for small to medium sized mammals for prey – birds can be more difficult to catch. In parts of North America, jack-rabbits and ground squirrels form a large proportion of the diet, and in the French Pyrenees marmots form a considerable percentage.

In parts of Skye, there are rabbit warrens on the open hill and where their numbers are high, I have seen eagles repeatedly hunting over the same small area of hillside again and again, clearly watching for vulnerable rabbits. Sheep and lambs graze on the same hills, but I have never seen either attacked. If the rabbits were to be suddenly wiped out, then the eagles would have to find an immediate replacement food. And there are few hares or grouse there. On the adjacent mainland in Glen Elg, in an area which is isolated by high passes, deep sea lochs and dense forestry plantations, an independent population of rabbits was wiped out by myxomatosis. Eagles were accused by the local people to be feeding on lambs and subsequently proved to be so doing by a friend and colleague, Alan Leitch, who carried out a study on behalf of the Nature Conservancy Council. He found that the birds had been feeding largely on rabbits with occasional carrion (mainly sheep) during the winter. When spring came and there were neither rabbits nor carrion, the eagles, a pair which held the local territory, did turn to killing small lambs for food. The hunger-stressed female who was incubating was particularly guilty.

There are remains of lambs at many west Highland eagle eyries and roosts, but I have never been able to prove whether they had been taken as carrion or killed by the birds. On several occasions I've watched eagles hunting over hillsides where there were sheep grazing with lambs. Every time the ewes have brought their lambs near and that in itself seems to have been enough to deter the eagles' attentions. Likewise, I have seen remains of red deer calves at both western and eastern eyries, but mainly at those in the west. Red deer calves are too large for an eagle to carry to the eyrie in one piece and so the individual legs which are found on the nests or plucking posts have always been insufficient for determining whether they had been killed by the eagles or not. There are documented accounts of eagles killing deer calves, and a newly born red deer is certainly within the prey size range for an eagle. Especially as they lie weak and helpless hiding in the heather for the first few days before following on behind their mothers.

The only occasion that I have seen an eagle attack a red deer calf, or any red deer, involved a pair of golden eagles harassing a group of two hinds and two well-grown calves in March. Diving repeatedly, both birds appeared to be trying to separate the smaller calf from its mother which was running around in panic, kicking up with her rear legs as the eagles swooped low. The company disappeared out of view behind a ridge and weren't seen again. I did check the area later for any evidence of a calf being killed by an eagle but found nothing.

Most pairs of eagles that I have seen hunting together have also been in the pre-laying days of February and March. And larger family parties of three, four or even five birds were all seen at the end of the summer after the young had fledged and were learning to hunt for themselves. When hunting together, eagles follow a regular pattern. They work their way slowly along a hillside. One bird flies low, perhaps twenty metres above the ground, the other follows on at three or four times that height. By shadowing the lower bird, the top bird watches its mate's every move as it glides back and forth along the face. If the lower bird sees something it will mark time, then suddenly pounce. It might strike a kill first time, but more often the prey will escape, be it a red grouse exploding in rapid low flight over the moor or a hare zigzagging down the hill. That is when the second bird strikes. It has the advantage of height to gain greater speed and momentum in its dive. It can also see the prey clearly, and its likely path. An open target is more easily struck. If the prey is lucky enough to escape a second attack, the momentum of the second diving bird can give it extra speed with which to chase the prey in level flight. I have watched an eagle gain on and catch a grouse in such a chase. And in another chase an eagle gained on a ptarmigan in level flight after it was flushed by its mate – although I never saw the final outcome.

Eagles are well known to be able to dive spectacularly, that view of an eagle chasing a ptarmigan was much more athletically inspiring. The ptarmigan's small white wings were beating fast and furiously at a speed only pursuit could incite. The eagle beat its wings quickly, yet with a leisurely style in the long steady confident stride of a middle distance runner gaining on an out-run sprinter on the last straight.

I have also watched a pair hunting when golden plover and meadow pipit flocks had newly returned to the Sutherland moors in March. The lower bird flew over the heather, barely above the stems at times. The second bird waited-on about thirty metres up behind. A bird was flushed, a pipit, and no sooner had it risen than it was thumped to the ground by the second bird. Together, they rolled and spilled down the hillside. The eagle then re-established itself upright, looked up, down and around, then mantled its kill under a shroud of black wings. A pipit is a small prey item for an eagle and within a few pecks, pulls and gulps it was eaten all up. The eagle stretched its feathers back into line, took off onto the wind again, and the two eagles joined in formation to continue quartering the rest of the hillside as an asymmetric pair.

After the eagles have laid their eggs, one of each pair must always attend and incubate them. From then on, for the remainder of the summer until the young are on the wing, most eagles are seen hunting alone. They are not so obvious when hunting then. And they don't soar high in the sky so often, as they cruise low and inconspicuously over the moors. Solitary hunting eagles usually quarter the ground from a mid-height of about twenty metres, although they can hunt from high circles, it possibly depends on topography and weather conditions, especially the wind force. They can ride and master extremely strong winds. It is on calm windless days with no lift that their flight is laboured and heavy. Eagles are patient hunters.

Eagles glide with minimum effort, using the wind eddies to lift and drive them over ridges or valleys, and in autumn they often flush out packs of grouse; red, black or grey ahead of them. But they seldom bother to chase those birds as they are usually in a tight flock. There can be over two hundred red grouse clumped together and it would require considerable effort and technique to catch one from such a tight knot. Eagles seem to prefer to hunt solitary sitting targets which are either too slow off the mark, or have not even seen the eagle approach. Opportunism is the name of the game.

Once when up in a high corrie in the north-west, where golden eagles have a wide diet, adding such items as small rodents and snakes to their plate, my attention was drawn by a pair of hooded crows attacking and alarm-calling at something out of my sight in a gully. At first I thought that the hoodies were the cause of the mischief, attacking the nest of some smaller bird as so often they do. However, the tide had turned on them this time. A golden eagle swung out of the gully, the hoodies went crazy, frantically diving and chiding at the tail and back of the eagle. Then the eagle went back into the gully out of my sight. At that stage, I merely thought that the hoodies were mobbing the eagle while it ate something it had killed and carried to the cliff to pluck. I ran around the head of the corrie to gain a better view of the gully. Then I set up the telescope and watched the commotion. The eagle flew back into the gully again, swung up and lunged under a small lip of rock. It didn't land - there didn't seem to be enough room to land - and it stretched out a talon to grab something out of the corner. Then it looped back and onto a ledge on the opposite side of the gully. It quickly mantled a dark grey body, ripped it up and ate it with amazing ease and speed. Unsatisfied, it gulped and straightened itself out before swooping

An eagle's eye-view of a day-old red deer calf on a grassy hillside.
The calf's camouflage is good, although its shape is easily discerned, as its speckled dark coat is better suited to dappled light in woodland. Eagles do take young calves, but they should be safe as long as they lie still. Movement is the greatest betrayal of camouflage.

A mountain leveret lies low and still on a high mountain plateau with scant vegetation. It peeps through blades of stiff sedge, watching, ready to run off at high speed as a second line of defence. Leverets are perfect sized prey for an eagle to catch.

back once more to repeat the process. All that time the hoodies were jumping around it, flapping hard at its rear, pecking at its back and tail. The eagle simply ignored them.

Eventually, the eagle flew off to settle on a mossy slab down where the burn left the corrie. The hoodies continued to mob it. It continued to ignore them. They went back to fly in and out of the gully and I went over to investigate the scene of the crime. There in the wall of the gully, set in a cleft was the hoodies' nest. The wool lining was turned out; some of it lay on the bed of the gully below. The nest lay quiet and bedraggled. The hoodies sneaked away with a few croaks as I approached. Then all was quiet.

It was May and other hoodie nests in the area contained half-grown young, of a size easily swallowed in one gulp by an eagle. Set under a small overhang on a tiny ledge, the nest had offered protection to the chicks from most predators, although not from the eagle. That eagle had to make repeated assaults on the nest, and it had to twist and stretch while in flight to pull out the nestlings one by one. There was clearly no room for such a large bird to settle on the nest, and it was well guarded by the parents. The eagle had been hungry, and as food was scarce in that part of the country it was all worth the effort.

Why are hooded crows so marked – grey and black? I came across one striking example of the effectiveness of this unusual camouflage one spring day in Argyll. The leaves on the rowans were not yet out and this is a tree often used by crows to build their nests in. I was walking up a burn-side where there were the usual scattered remnants of Highland woodland. A bulky stick nest in the mesh of branches was silhouetted against the sky. As I climbed higher uphill and passed by on the bank less than ten metres above, I looked down to see if the nest was occupied. Well, I could hardly believe it; a hoodie was sitting tight on her eggs. And where the bare top branches cast shadows across the lower branches, there was a grey and black pattern. I never noticed the sitting bird until I looked straight down on her, so good was her cryptic colouring. Of course, as soon as I focused my eyes on hers she was off, she knew she had been detected. Her grey was the same as the grey of the bare branches and the black was that of the shadows. The shape of a pure black carrion crow would have been easily identified in such a situation, and fall prey to an eagle.

Golden eagles are successful opportunists and their technique is not restricted to species. This extends to the types of land they will hunt over. Given the space and left unmolested, eagles will hunt sea cliffs, open woodland, and low moorland, as well as the hillsides, corries, high ridges and plateaux. When an eagle cruises along below the lip of a sea cliff on the west coast or out in the Hebrides, great curls of fulmars spill from the ledges, their white spray foaming all around the steady dark wings of the hunter. The fulmars rely on their great numbers to confuse and disturb the eagle, although its attention seems so fixed and oblivious to their commotion that it is inevitable that one of their number will be caught – one that is slow to launch off its ledge.

Eagles need to take care when hunting fulmars however, for they have one effective and potentially lethal defence. They spit. Or rather they can project vomit, a foul-smelling stomach oil, which clings to feathers, matting them and crippling a bird's flight ability. If an eagle were to miss-time its attack it could find itself smeared, flightless and would starve.

On the high tops, the ptarmigan rely on non –aggressive defence. They initially clamp down low when an eagle sweeps overhead, relying on their cryptically coloured plumage and stillness to conceal them from the eagle's alert eyes. If they are caught unawares, and chased, they quickly form tight packs and race hard. They usually fly around the upper slopes of the hill they are on, or occasionally they will flee across to the next hill. They know their home ground well and will head for the boulder fields and land in a hard breaking action, almost crashing into, and scurrying under, a patch of jagged rocks. There they will be safe as eagles are too cautious to dive in after them. They can't risk breaking a wing or a leg on the stones. And the chance of an eagle being able to reach them under the rocks is slim as I've seen.

Hunting requires good vision, stealth, speed, strength and agility; the golden eagle has them all and its versatility and prowess perpetuate its reputation as a masterful hunter. People are overawed by all this and invoke miraculous powers. Eagles: they know they have limits.

A commonly portrayed image of an eagle making a kill is of a bird diving from high in the sky, grabbing an animal and immediately flying off with it hanging from its talons. But eagles, like all raptors, seldom kill like that. Golden eagles mostly kill prey smaller than themselves by a sudden stoop from low in the sky. They fold their talons around the chest of their prey and hold tension until they feel the prey stopped breathing, or at least struggling to get free. Most prey probably die of suffocation from constriction, although the eagle will nip at the back of the head and this could account for some prey, particularly larger animals. When they attack very large animals, such as deer calves, they will grab at the head in the first place. They probably would know they could not restrain them any other way, and they likely kill them with a fatal stab with their talons. Once the prey is dead they will often pull it to a convenient place where they can pluck it and eat in peace from scavengers' attentions. If they are feeding young, they will carry it away to a plucking post, such as a hummock or branch near the nest where they will prepare it before presenting it to their chicks.

A cock ptarmigan sits high amidst rocks watching over his territory and mate while she sits secluded on her nest and eggs in the heath below. Whenever danger in the form of a potential ground predator is seen approaching, he will give a loud croaking alarm call and fly off to distract the predator from the nest. When an eagle approaches, he will tilt his head sideways to watch it and slowly crouch down. His best defence against aerial predators is his intricate cryptic plumage, and his close proximity to small rounded rocks which are similar in size and shape to himself. Meanwhile, his mate will sit low and motionless on her nest until the eagle has passed.

A hen ptarmigan crouches low on her nest. In spring and summer, her plumage is different from that of the cock. The feathers on her head, back, tail and wings are a mix of russet earthy tones which match the dark colour of the mosses, the young blaeberry leaves and the shadows beneath the grasses. And the curved buff tips to her feathers blend in with the pale curling blades of the previous year's grass stems. She is very difficult to see - even for an eagle.

Capercaillie are precariously close to extinction in Scotland. They were lost once before, most likely as a result of clearing native pinewoods and shooting. Then a viable population was subsequently re-established with birds imported from Sweden in the 1830s, and the later prolific development of conifer plantations probably helped. Now, their demise is due to such causes as full-grown birds being killed by flying into deer fences within and adjacent to pinewoods, and chicks failing to survive in the recent wet summers. Capercaillie are long-lived birds, and so, as long as people see them around, they believe the species is not in danger. However, the population's future depends on the old birds living long enough to replace themselves with young reared in favourably dry years. Forestry managers are now trying to reverse the decline with more favourable forest design.

To lose a species once may be regarded as a misfortune; to lose it twice would be carelessness.

4

CAPERCAILLIE

Weighing in at four kilograms, a cock capercaillie is a somewhat large bird for a golden eagle to kill. A female eagle weighs about five kilograms, a male about three and a half.

Green crowns of Scots pines linked one into another all along the hillside. Here and there the pale fresh green of birch trees added light to their darkness. Chaffinches' songs rang through the wood from one end to the other, and down at the edge, two roe deer bounced off into the depth, their grey-red backs so much like the grey-red bark of the pines. Skimming over the canopy was an eagle, a cock, and soaring a hundred metres above him was another eagle, his mate. The male slipped and weaved along the wooded hillside, turned and quartered another tier down below. Suddenly, he stooped. And his mate immediately dropped to fifty feet directly above him in one sudden swoop, ready to back up his attack. But they pulled out of the chase, and together they circled up on shallow upturned wings and continued their hunt.

A pair of buzzards mewed from a few hundred metres off and came into the scene with a rush. They dived at the eagles when they first approached, only turning aside at the last second. They called again, and a third buzzard appeared. All five birds were spinning and twisting knots in the air. The eagles turned effortlessly away from the buzzards' feeble strikes. One flip over and a flash of the male's talons were all that was necessary to shy off any close threat from the dwarf buzzards. Then, with a flick of their wings, the eagles were both quickly up above the buzzards and try as they might, the buzzards could not regain the advantage of height needed to strike down on the eagles. With their superior soaring power, the eagles were high in the sky in a minute or less, and the three buzzards slipped down into the trees, out of sight.

The scene appeared to be resolved when a third eagle entered. Judging by the amount of white under its wing and tail it was a young bird, from a nest of the previous summer. It was accepted by the pair without challenge or aggression, so perhaps it was there own offspring. All three continued to hunt together over the wood. Their turns and dips through the treetops repeatedly confused my perception of who was who, and where they were in the sky, but the adults had a pattern to their course, the youngster was the one with the random flight. Suddenly the pair both folded their wings and dived deep through the canopy. At my last glimpse, they had brought their talons forward from under their tails, although at the moment of strike they both disappeared from view. Out burst a thundering cock capercaillie. His head was well forward. His bill was wide open. His throat feathers were extended to spikes. They had missed.

The eagles rose out from the trees behind him. They caught the wind and continued to soar low over the wood in tight turns. Once more the male dived into the wood; this time the female held aloft, waiting-on. Out flew another cock capercaillie, closely followed by a hen, both like the first bird in full stretched alarm. The eagle had missed again, so he reclimbed the air. Down below, he had left a birling confusion. He dived a third time and another half dozen cock capercaillie and three or four hens slipped out through the trees. They dodged round the tight tops, not daring the dangerous open sky just above where the hen eagle was waiting to strike down any bird that gave it a chance.

The eagles had found a capercaillie lek and had been tempted to try for a kill. Although very large, and at the top of the size range possible for a golden eagle to kill, I have found capercaillie remains amongst prey items. It might have been possible for those birds to catch one on that occasion, especially by hunting together. But, it was not to be.

The pair of eagles soared on, still low over the wood, working over the undisturbed far end of the hillside. The third bird had played no part in the chase. It had only hung-on, high over the hunting scene, and when the excitement was over it soared away even higher and beyond the top of the hill. As it crossed the horizon a peregrine tiercel dived and stooped repeatedly onto its back. Casually, the eagle rolled over and fended it off with its talons and set off on a rigid, purposeful glide over to what I knew to be a nesting cliff of both peregrines and eagles. Meanwhile the pair were almost out of my sight, but still hunting, and being harassed by yet another raptor. This time it was a kestrel. It looked so tiny and optimistic compared to the buzzards and peregrine. Yet, its enthusiastic torment of the eagles succeeded where that of the larger birds had failed. The kestrel's persistence drove the eagles out of my view over the horizon, last seen with a spit-firing kestrel dog-fighting their tails.

Golden and white-tailed eagles both hunt over the islands off the west of Scotland, large and small, where there is a considerable overlap in their diets. Each takes fulmars from the sea cliffs and rabbits on the machair (the rich grasslands between the dunes and the peaty ground). The golden eagles prey more on hares and grouse on the moors; whereas the white-tailed eagles can also catch prey on the water and so extend their diet beyond that of the goldies. They can take fish that come to the surface, as well as ducks and auks. And they seem to specialise in snatching fish from otters as they eat their catch on the shore.

5

WHITE-TAILED SEA EAGLE

Since being persecuted to extinction at the turn of the twentieth century, the white-tailed sea eagle has been re-introduced to Scotland, and now at the turn of the twenty-first century, it might just have become successfully established as a breeding species once more..

When I first opened a map of the Outer Hebrides, I was amazed by the incredible number of lochs, lochans and sea-lochs. The whole map appeared to be blue. Haphazard blocks of land were separated and scattered in all directions by water. Irregular rings of brown contour lines appeared and disappeared. Some long and widely spaced ones meandered over the bogs, other close set loops strode up the hillsides. Dashed here and there were couplets and blocks of black wavy lines indicating only the steepest of the crags. The rest lay hidden where the contours ran close or overlapped one another. Apart from these natural features there were the thin red, orange and yellow road lines which linked the far flung communities, and there were the odd cartographers' hieroglyphics. Any other or more exact details of the islands' topography have to be filled in personally from experience.

With just such a map I set off into the knobbly landscape on a pre-set course. And it wasn't long before I made a deviation here or a little turn there. Only the top of the ridges and hillocks were dry, so wellington boots were essential for one who was following the passes. I had carefully plotted the best line to my destination and this followed a sequence of bearings which avoided the lochans. Although this pattern led to me wading through rank heather, moss or wet bog it proved to be easier at the end of the day than repeatedly climbing up and over the greasy crags which lie unmarked on the map and unwanted on a walk.

The going was heavy, but not as slow as I first feared it would be. I wove my way across the bogs, the unfolding views and aspects keeping my attention away from the trudge. There was a scar of burnt heather which I launched across with an idea of saving time. That was not the case however. The soil was so wet and thin that there were no footholds on the slimy, mossy peat and I slipped and slithered all the way across the bare patch. The fire had been too hot and unmanaged, and it had burned off the thin, precious layer of litter and topsoil. The odd sedge shot out through the peat creating a hint of greenery, but that was poor grazing for the sheep which the fire was meant to help. It would take years for the rich grasses and herbs to come back.

On one secluded lochan a pair of golden-eye ducks were fishing the dark water. Most of the lochans which lie on the gneiss bed-rock of the Hebrides are dubh lochans, that is 'dark little lakes', so called because their waters are acidic and peaty which makes them, well, dark. Very little life, either plant or animal, can survive in these waters compared to the rich variety that thrives in the lochs which lie on the more fertile land of the machair along the western shore of the isles.

As I came near to the cliff side of a long sea loch I saw a pair of golden eagles circling ahead. They flew low over the moor, one above the other, and intent on hunting, they were indifferent to my approach, merely jinking to one side as I walked over land which they had just quartered. Then they carried on hunting the moor behind, which I had just crossed. The birds were still in my view when I reached the shore of the loch, or rather the top of the cliff which dropped sheer into the cold colourless water. After a long, comparatively empty moorland walk I was suddenly thrust into a rich, thriving community of plants and birds. The scent of the sea air blew up over me, bringing with it the flavour of thrift, scurvy-grass, stone-crop and sorrel. Bright yellow lichens crusted the ribs of the rocks, ivy swarmed across a wall and an ancient holly twisted tight into a corner.

Ravens and hooded crows cursed my arrival and a scouting great black-backed gull skimmed along the cliff, its cold eye assessing me closely as it passed by. Across on the far side of the loch, the gloomy north-facing aspect, the wet green face of a broken crag was dotted with fulmars sitting on their nest ledges. They gaggled and gurgled as birds flitted to-and-fro between the ledges, selecting and guarding their own little nest scrapes. It would be months before they laid their eggs, but the prime nest sites are valuable real estate and need to be occupied early to ensure their later possession. Endless circles of fulmars swirled around the cliff-face, and some also flew along my cliff, only a few metres away, blowing what sounded like a kiss as they past me.

The scene was hypnotic as I lay in the sun. Tucked deep into a bank of rank heather, I was sheltered from the cold showers which blew over my shoulder, fighting the oncoming spring. I was lying there in comfort, slowly scanning the cliffs through the telescope, searching for eyries, when a pair of hooded crows snapped me out of my trance. They were jumping and hopping about on the cliff, their grey bodies blinking behind their flapping black wings. On turning up the magnification of the telescope a third body appeared in the melee. A grey head with a heavy sulphur bill bobbed at the edge of a woodrush filled ledge. I sat up and I held the telescope firmly at maximum power. It was the head of a white-tailed sea eagle jerking up and down as it plucked the feathers from a dead fulmar. Low set between massive folded wings, the eagle held a sharp deliberate pose, a perfect exhibition of power and strength. As the bill dug into the carcass, grey and white feathers trailed off in the updraft, spreading themselves far and high across the cliff. Intent on its meal, the eagle tossed away some blood stained, sticky red feathers over its right shoulder and gulped down the flesh, with short choking gulps.

Together with the hoodies, I watched every move that the eagle made for the hour and a half that it took to complete its meal. I was amazed, and impressed, by the nerve and impertinence which the two hoodies showed. They poked and tweaked at the fulmar while it was still held firmly under the eagle's massive talons. Never missing a chance, they would run or bounce around behind the big bird to pick up a scrap. Perhaps a feather with some skin attached but nothing much more. Only once did the eagle take a swing with one foot and a wing at the scroungers, but that was when one of hoodies was all but pecking at the fulmar in turn with the eagle. Under no circumstances did the eagle take both feet off its prey. It was well aware that one of the pair was ready to jump in from behind and steal its food if it attacked the other. Eventually, the hoodies' persistence was rewarded. The eagle dropped off the ledge and flew across to another some fifty metres away. They pounced on the leavings scattered over the plucking position. Meanwhile, all chaos broke out as the killer took wing. Alarm calls rang out from fulmars and gulls as they clamoured away from its path. Tripping off ledges like pearls from a string. Eventually the panic passed once the eagle re-settled, and the ranks were reformed.

On its new perch, the eagle was now in perfect view, and it immediately set about a full toilet after its long meal. Straightening out its wings in a long slow stretch it raised its tail feathers and sent a great white splash far down the cliff face. Then it flicked its wings closed and resettled its feathers. Taking one wing at a time, it proceeded to meticulously preen every primary, secondary and tertiary feather. The tail, when its turn came, was given more attention than

Golden eagles live at high densities in parts of the western Highlands and Islands where there are large amounts of sheep carrion on the open hill. However, there is generally less live prey, e.g. mountain hares and grouse; and fresh meat seems to be important for fledging success. So, although the adult population is more dense where there is carrion, the average number of chicks reared is lower than that in the areas where there is plentiful live prey, such as in the eastern Highlands. Of course there is not a strict east-west divide as there are some home ranges in the west that contain good stocks of live prey, and some in the east which do not. A number of eagles live near seabird colonies, others live where mountain hares are abundant, and rabbits are plentiful in numerous local pockets. Eagles in such places in the west can breed well.

Land use practices in the Highlands are always changing and have led to less sheep on the hills and less deer in some areas. It is likely there will be further differences in the future and only long term study of the eagles will tell if there is any effect, better or worse for the eagles.

either wing though it held much less area, and the breast and belly feathers were also adjusted. This demanded some awkward manoeuvres by the bird as it twisted its head under and around the large wings, and all that delicate preening was carried out with the same sharp bill that had just butchered the fulmar. The bill itself was cleansed with a wipe on a cushion of woodrush and picked precisely with a long black talon toothpick. To finish, the eagle stretched itself upright, shook all its feathers in a labrador swirl and sank back into a squat, huddled posture with its bill resting deep under its back's feathers.

For an hour or more all was at peace from end to end of the loch. The sea was calm in the shelter of the fjord setting. Out beyond the magnificent headland an occasional squall passed by to ruffle up a little white water. Between the showers, a shimmer of glistening water ran along the length of the loch from the ruins of a croft at the head out and lost by diffusion over the empty waves. I lay back in the heather and relaxed after the continuous excitement of watching the eagle. I took the chance to drink a warming cup of tea from a flask and looked out over the Minch to the other islands and the mainland beyond. There was snow on most of the higher hills and while picking out each mountain profile in turn from north to south I counted the other occasions when I had seen white-tailed eagles in those hills.

With young birds dispersing all around the country in search of territories, and despite their name, sea eagles might be encountered almost anywhere in the Highlands. Every eagle should be checked as to whether it is golden or white-tailed. On my first meeting with this bird I saw two together. That was in the late seventies on a sea cliff on Rum. They were still only young birds and so, unfortunately, did not have the adult white tails. The tail and head become pale when they are about five years old and mature enough to breed. I have seen other young birds far inland. One was crossing a high pass over the water between east and west coast rivers. Another was as far from the sea as one can reach in Scotland. That sighting was in November and the bird may have either found a source of food in the grallochs left behind on the hill by deer stalkers or it might have been feeding on the kelts below the salmon spawning beds.

Whether young or old, the silhouette of a soaring sea eagle is a grand sight. The wings are both broader and deeper than those of a golden eagle, the tail shorter and wedge-shaped, and the bill heavier. Once it is aloft, the sea eagle holds its wings stiff and straight. The golden eagle holds its wings in a shallow V-shape. And it is while soaring that the sea eagle takes on a more vulturine than aquiline pose. I have watched one hang high on the breeze for over an hour, its head slung low, scanning the ground far below. Another pair of birds which I remember soared in this style regularly over a favourite ridge. Occasionally one would dive into a display of deep undulating flight. Or both would display together. When they landed on an exposed rocky knoll they made deep bows and lengthened their necks as they yelped to one another. They are much more vocal birds than the golden eagle.

I have only seen one kill by a sea eagle, and all I saw was a bird dive into a hillside and rise up with a rabbit hanging from its talons. Close on the heels of the eagle were a pair of buzzards. Both species kill rabbits and both species are carrion eaters that can scavenge each other's kills. On that occasion, whether the eagle or the buzzards killed the rabbit I don't know, but the sea eagle won the prize and kept it.

Eagles, like all birds of prey regurgitate pellets - balls of fur, feather, bones and other large indigestible food items. In the numerous pellets I have found cast by sea eagles, the most intriguing prey remains have been the fur, whiskers and claws of an otter cub. Sea eagles are known to snatch prey from otters, but that bird must have snatched more than the prey. Fish bones might be expected to be the most common remains, however, either these soft bones are easily digested and are rarely cast up, or as fish stocks in our seas are so low, the eagles might seldom eat them. The most common of the prey remains that I have found are the bones and feathers of seabirds, especially those of fulmars. Fulmars are now omnipresent around our shores, but when the sea eagle last flew our skies at the turn of the century, the fulmar was an unfamiliar bird. For centuries it was only known to breed on the far outlying islands of St. Kilda. It would be a wonderful achievement if the re-introduction programme resulted in the sea eagle becoming as widely distributed as the fulmar has become.

Thinking of fulmars, I looked out over the loch to the bird of the moment. I re-adjusted the telescope and continued to watch the eagle more closely. All was quiet for several more minutes. Then with a surprisingly quick decision, the grey head lifted once more and in a short, quick drop the sea eagle was about once again. Long, straight fixed wings soon lifted the bird back up on the updraft. It glided dead on line for me. In seconds, it was well out from the cliff with a cloud of fulmars, gulls, ravens and hoodies, all nagging and squealing in the utmost alarm. The eagle didn't heed any of this, it simply glided towards me. I soon dropped the telescope as the bird was travelling much too quickly to follow in the tight field of view. I picked up my binoculars, but even through those, I found it difficult to keep the bird either in view or in focus. My neck was cricked, so I dropped the binoculars, and there was the eagle right above me. Within about twenty seconds that bird had glided right across the loch without a single flap of a wing since leaving its perch.

A minute previously I had been straining my eyes to see it through a telescope set on a times-sixty magnification. Now I was straining my neck to see it thirty feet above me. Judging by its behaviour I don't believe that it had seen me, although by the line of flight I had initially thought so. When I dropped the binoculars and showed my pale face, the eagle arrested its glide with a reverse

Carrion forms a large part of the golden eagle diet in the Highlands. Here a red deer hind has died while giving birth to a calf. The calf's leg can be seen projecting through the pelvic bones and its tiny ribcage lies above the hind's lower vertebrae. The flesh has been most likely eaten by eagles, for the bones have not been chewed down as they would be if a fox had eaten it. Maggots would have further reduced the carcass.

Fulmars are a common prey item for some eagles in the north-western Highlands and islands. They seem an ideal food source as they are about rabbit size and slow on land. However, they have immediate and long-term dangers. They can accurately spray any potential predator with projectile vomit of sticky, fishy oil; enough to ground and subsequently cause the death of a raptor from starvation. Some eagles do obviously perfect a method of safely catching them, but they risk a build up of toxins in their bodies, for fulmars, like most marine life, are themselves potential carriers of poisonous chemicals, such as polychlorinated biphenyls and mercury. The concentrations in the eagles might not be enough to kill them but perhaps impair their reproduction.

swirl of its wings. The span and depth of those wings were immense when seen from so close. The freckled white tail by which the bird is named spread to help break the glide and for a moment the bird appeared to stall. Immediately above me, it wheeled around and soared slowly up and up in the tightest of turns that its wingspan would allow. I looked at it and it looked at me. The exaggerated bill which had first caught my attention did so again. My focus jumped from that to its eye and along to its yellow feet tucked into the soft down below its translucent tail.

Instinctively, I checked its unfeathered tarsi for rings. By its age and the year, this was one of the birds released on the island of Rum; it would have had a single metal ring and a unique combination of coloured plastic rings from which I could identify the exact individual. Unfortunately I couldn't see any rings. So I picked up my binoculars again as the bird rose up higher. It could have had a grey metal ring hidden by its down although it definitely did not have any bright coloured plastic ones. This could well be expected, for the bird was in full adult plumage. It could easily have lost or picked off the plastic rings during the five or six years of its freedom since leaving the feeding station on Rum. In later years fledglings have been fitted with soft plastic wing tags, with individual colours and numbers, letters or symbols, so the birds can be much more readily identified.

I was lying on that sea-girt cliff in wonder, watching the sea eagle soar up and up, when it unexpectedly flicked to one side. A golden eagle had dived at it. From behind me, the pair of golden eagles I had passed on my way over the moor had come in to mob this new-comer which threatened to invade their territory. Both goldies joined into the thermal that the sea eagle was climbing. And that was when I really appreciated, for the first time, the size difference between the two species. The vulturine features of broad straight wings, stumpy wedged tail and long bill of the sea eagle dwarfed the slightly bent wings, long tail and slighter bill of the goldies. All this while, I was sitting directly below the three of them and it was difficult to see how anyone could ever mistake one for the other as they were so obviously different when seen side by side. But those features are relative and vague when seen at a distance with no other bird for comparison, as is the usual situation when one or other of these birds is seen in the Highlands. And I have had difficulty telling either species for sure several times since, usually young birds at a considerable distance.

The sea eagle continued to soar up. It ignored its tormentors who were being left farther and farther behind as the broader wing area lifted it percep-tively more quickly. Once high enough for its purpose, the white-tailed eagle cast off and glided back across the loch. The male goldie took off after it, the female made a half-hearted attempt before turning back. Flapping hard, the goldie caught up with the big bird - golden eagles can fly deceptively quickly in level flight. From below and behind, it swung up and twisted to strike the white tail feathers. Undeterred, the great bird slipped farther away on the same steady glide, and the goldie lost its stride, but kept the chase for a while before accepting defeat. Or, perhaps it had won, and was flying back to join its mate after successfully driving off the competitor.

One steady line of a glide led the sea eagle over a distant ridge with no height to spare. The bird had measured its initial take off height from the thermal with absolute perfection. I turned away and zigzagged across the moor with much less precision. And on my way I saw the golden eagles back together, soaring around a small hill on which they had built up one of their eyries. Somewhere, sometime, that wandering sea eagle might have found a mate and built a nest too.

A few years later sea eagles began to nest in the glen which the bird described above had flown into.

Since the white-tailed eagle became extinct here in 1918 after a long familiar tale of persecution and collection of skins and eggs, mostly during the previous century, it is heartening to know that homebred sea eagles are once again flying around Scotland. Between 1975 and 1985 eighty-two white-tailed eagles were released on the island of Rum in an attempt to re-establish a native breeding population of this magnificent bird in Scotland as part of a re-introduction programme run by the Nature Conservancy Council (now Scottish Natural heritage) and the Royal Society for the Protection of Birds,. The new stock was nurtured from chicks collected in Norway, and most of the early fieldwork was carried out by one man, John Love. In 1985 a pair bred successfully and the next generation of Scottish white-tailed eagles was begun. Then in 1995, the first Scottish bred chick began to breed. This success has since led to 52 pairs rearing 42 young in 2010. A further 58 young birds were introduced to the northwest Highlands between 1993 and 1998 and there are now birds spread all over the Hebrides. Since then a third phase is in progress to release a further 100 chicks in east Scotland between 2007 and 2012. So now, a Sea Eagle Project Team of professional and voluntary workers is required to continue monitoring their progress. Unfortunately, in recent times, white-tailed eagles are again being persecuted – three have been found poisoned in the past three years. Clearly, there are still landowners, gamekeepers or shepherds who still hold the nineteenth century attitude towards raptors which resulted in the extermination of the original stock. How many others have been killed but not found?

By sunrise, eagles have usually slipped out of their roosts and are perched on a high vantage point to watch for prey. Grouse, hares, rabbits and most other potential prey species are especially active at dawn (and dusk). Eagles are aware of this and take advantage of these activity patterns.

6

A STILL DAWN

When there is no wind, flying demands considerable effort and our largest raptors, the eagles, are seldom seen in such conditions.

The grey dawn hung on to become a full grey morning. No air stirred and smoke rose straight from the chimneys on the string of cottages running between the fields and the woods.

A woodpecker's drumming rang through the glen and bubbling blackcock could be heard sparring with one another on their lek, an area of short deer-grazed blaeberry and grass where a wood had been clear-felled. The stumps were used as perches by the ring of young cocks watching and waiting for their turn to joist. Meanwhile, out on the deep heather of the moor, two cock red grouse strutted up and down, side by side, pumping their scarlet combs fully erect and puffing out their breasts. They were defending their territories for the approaching nesting season.

A pair of kestrels sat on the crimson limbs of a tall birch tree not yet in leaf. The rich russet brown on their backs, especially that of the brighter male, showed up well in the low spring light. Together, they sat with their heads down, scrutinising the grass bents for the twitch of a vole. Down by the riverside, a wee cock merlin flew up, carrying a small brown bird in his talons, trailing under his tail. His prey was most likely a meadow pipit for they were the only numerous passerines about in the area. Most of the summer migrants had not yet

A mountain lochan shines bronze in morning sunshine - the view from an eagle cliff.

What a start to another eagle day!

arrived. A curlew jumped into the air in chase of the merlin crossing its territory on the way up to the hillside above. Its loud whaupping alarm call sliced through the glen and the merlin jinked away with its booty, as if hurrying to get away from the chiding scold.

Across on the opposite side of the glen, a pair of peregrines were mobbing a pair of buzzards. Their raucous screams echoed far and wide out from the cliff face. Although there were several raptors about, none of them was displaying. Hunting on a windless day demands a lot of energy, high flying displays must demand even more. Then a solitary eagle flapped tediously along the ridge to land on a branch jutting out clear from the top crown of a pine tree. From there on its high perch, it began to hunt, turning its head this way and that, ignoring a group of three blackcock which flew away in panic as it approached. They were too fast and strong, the eagle would have preferred a rabbit or hare sitting quietly in an open patch of heath.

Further down the hillside there was a buzzard sitting atop another white limb of a scots pine, one that stuck up above the forest canopy. In the flat light the buzzard looked as big as the eagle, but it lacked the sturdy head and shoulders of its superior. The eagle sat bold on the skyline. The buzzard sat low in the glen. Tree tops are a favourite vantage point on such breathless days. And a cock song thrush sat on another dead branch of a pine. Dead branches usually dry before the leafy ones and many species of birds prefer them on dank days. The thrush blasted out the repetitive string of notes in his song, and I was surprised to hear his call rather than that of a mistle thrush which is more typical of the mixed pine and birch woods. There had been some mistle thrushes singing earlier, but it was now late in the morning and still early in the year, so most other birds had stopped singing. Only the descending trill of a chaffinch rose from the wood and, in duet, the songs resounded on through the forest again and again.

A light mist crept stealthily into the glen. The buzzard left its perch with a few low swoops, and the eagle slipped away without me noticing when or where to. Both trees were empty. The wind lifted with the day, the song thrush was gone and so was his song.

Early morning mist lifts off the calm waters of Lochan na h-Achlaise, Rannoch Moor. On mornings such as this, eagles will glide from their roost to perch on an elevated hillock, a cliff or a bare tree branch. They then will spend hours just watching and watching, waiting for potential prey to expose itself on the moorland below. If there is no prey, they will probably not fly until the air stirs, the mist lifts, and they too can lift on a breeze.

In over thirty years of eagle study I have never seen a golden eagle perch on a telegraph pole, not even by the remotest of roads. I would not say they never do so, all I can say is that I have seen hundreds of buzzards on poles.

When the clouds roll into the hills they reveal the pattern of the winds which we could never otherwise truly comprehend. Eagles can feel the eddies swirl as they fly and casually adjust their flight feathers to ride the turbulent skies.

7

EQUINOX

Spring weather in Scotland can be stormy, but life must be passed on.

Five days before the vernal equinox, gales swept the tops and rains washed the slopes of Sutherland's hills. Hoodie crows scavenged the shores at the sea-loch's low-tide. And grey winter rowans bent their crowns low in the lee of the river's banks. Four days before, low mist and showers, driven hard by the wind, rolled over the moors. A pair of golden plovers peeped from somewhere unseen, their gold-speckled backs lost in the mist-speckled heath. Three days before, the morning mists lifted and the clouds broke for a while. Oystercatchers brought the dawn in with a shrill piping fanfare. Lapwings twisted black and white loops in the air, and curlews rose high over the moors exhalting the arrival of spring. Far away on the skyline, a faint speck slowly soared around and around. An eagle was back in the sky.

Farther on, in an adjacent glen, another eagle soared over its territory too. The wind held steady in the north-west and the bird hung like a mobile on the cool aerial stream, twitching a wing or turning its tail to keep itself fixed over one spot. When along came a pair of ravens chiding and mobbing the eagle. They cut the string with their tormenting and the eagle dropped down into a cliff.

Black clouds rolled on, keeping high, building themselves thicker and heavier until they filled the whole sky, and the hills turned back to their gloom.

The cold wind wiped the sun from their face, and although it kept fair for a few more hours, spring had lost its hold and winter had struck back. A triangle of pillars climbed up from the moor to a point on the ceiling of cloud, and were topped by a jumbled up pinnacle of colourless, shadowless blocks. A pack of ptarmigan scattered along the top edge of the cliff, each bird burring its wings at full speed and holding its head hard forward. In a panic, they flew as a flock, but kept an irregular rank, and as they rounded a buttress they dropped all as one into the cliff and out of my view. Into my view came the wide wings of an eagle, then the wings of its mate. Hunting together, they were flushing the ptarmigan from the summit screes, waiting for their chance to pick off any one which was slow to take off. The eagles cruised along, up and over the line of the pillars and back again for another sweep. One bird kept close to the ground, frightening ptarmigan out from the nooks in the mountainside. The other bird slipped along behind its mate on the line of the hill crest. They flushed a score or more ptarmigan, but never tried to chase any.

In the height of the confusion of harrying eagles and rattling ptarmigan, a pair of peregrines swept by to add their voice. They sliced through the paths of the eagles, stooped and rose back up on the wind at neck-breaking speed. Ignored by the eagles, they shot out of the scene. Had they stooped at the eagles? Or had they been trying to catch the ptarmigan flushed by the eagles?

Trails of mist dropped from the summit, and rain streaked the head of the glen with grey and dark grey. No more ptarmigan showed on the cliff, a cliff now glistening black and white where water began washing the slabs. I scanned around for the eagles again and was struck by the heraldic profile of an eagle perched on a pinnacle's very point. Ten minutes later the other eagle landed close by. As the first bird was the larger of the pair, I took it to be the female. Both were in splendid condition with full compliments of flight feathers. She was the darker, and he could be distinguished by his gold epaulettes. Together, they fetched a fine pair of gothic gargoyles.

Ruffling his wings and shaking his head, the male looked across to the female. She sat firm, head set into the wind. With exaggerated sideways strides he ambled up to her side and gave a slow bow. In response, the female turned away to the side and held a long bow with her head stretched low to the fore. The male immediately jumped onto her back. She lifted her tail aside and held firm in position. He kept himself steady during gusts of wind by half opening his wings and shuffling his legs. For, while on her back he had to support his weight on his 'heels' with his talons folded closed. Thirty seconds later their coition was complete and he jumped down to her side. Each bird stretched, preened, and sat side by side for a moment. All the while, veils of mist rolled on by, spraying the pair with beads of water. The birds in turn sprayed off the water with a toss of their heads and a spin of their feathers. Underneath an eagle's tight top layer of waterproof feathers there is a thick wad of white down, and they would be warm as toast inside.

After five minutes the birds began to shuffle about on the pillar. Perhaps they could feel a worsening of the weather approaching? It was time to move on and the male stepped onto the updraft and followed it to the thick edge of the mist. From there, he dived down past his mate to the foot of the cliff and rebounded back up to the ceiling. He did seven repeats up and down, showing off his elastic control of the air to his mate who sat braced against the strengthening wind throughout the display. When he had finished and hung motionless on the updraft she took to the wing, and in one fading glide she slipped into the mirk and out of my view. I looked back up for the male, but he had gone too. Mist curled around the pillars, and rainwater ran down the gullies. I should have been gone too – long gone. It was cold.

The clouds had reclaimed their hold on the hills and a storm raged around the tops. Down below, the moor and glen looked the bleakest of bleak. Under the pillared cliff there lay a foundation platform of a lower cliff system, and tucked safely under an overhang of that low cliff there was an eagles' eyrie. I spied it through the telescope, in the very very poor light, and I could just make out that it was built up with freshly gathered sticks and heather, and the lining was filled deep with woodrush and grasses. All was ready for eggs.

Two days before the equinox it was clear but still windy, and I didn't see an eagle all day. A fox was the day's highlight as it bolted from a cairn as I climbed past. One day before the equinox, heavy hail showers came on a strengthening gale. It was clear, almost sunny in between the showers, and the one eagle which I saw hunting retreated to the shelter of a corner in a cliff when a hail shower edged towards it. Not long after, it was dusk and still it had not re-emerged. It had probably gone to roost for the night. On the equinox itself, the hail turned to snow and storms thrashed the land. I watched one pair of eagles sit by their eyrie from morning to dusk. The male remained on his perch all the while I watched. His mate moved once, and only once, to pick up a piece of a ptarmigan from the edge of the roost ledge and eat it alone. Nothing else happened and I left as night fell on the two huddled up figures quietly waiting for fair weather in which to get on with their lives.

A large yet still modest scale golden eagle eyrie. Some nests, particularly those set in Scots pines can become very large after years of use. Most of the branches used in their construction are about an inch in diameter. Nests which are set on cliffs far from any substantial woodland are built with long strands of heather and seldom attain great size. Often they are blown off the nest ledge once abandoned by the birds for the year. (Ewan Weston, seen here climbing to the nest, is over six foot tall and he is forty feet up the tree).

8

NEST BUILDING

Golden eagles build their eyries in trees or cliffs and some of the cliff nest sites have probably been used for hundreds if not thousands of years. Many of the ledges have been selected by subsequent generations of birds, and there are old place names that relate to where eagles have historically nested, and still do so today.

A rustle of feet on the frozen sponges of moss and frosted leaf litter made me fix still. There wasn't a breath of wind to carry my scent, and the grey frosty air from the previous night was hanging motionless over the gaps in the trees. I could see nothing at first and the footsteps halted just at the same time as I did. Somewhere in the scrub a pair of eyes was watching now that I had stopped moving. But the eyes never saw me, or at least they never headed, for a stoat came louping out of the bog-myrtle.

The stoat's fur was pure regal ermine, glistening white except the last inch of black hair at the tip of its tail. In Scotland, stoats do not always turn white in winter, and it is interesting to note when and where. I met this one in March in the eastern Highlands at about two hundred metres above sea-level, the habitat was mixed scots pine and birch woodland. I had seen another only the previous day on a loch shore by the sea, working through rough pasture and reeds, and that one was brown with not a hint of white. Their colour change is triggered by a hormonal response to temperature which in turn varies with altitude, latitude, and in the mix of temperate and sub-continental climates that occur across the country. So, the stoats in the colder, higher and more northern areas are more likely to turn white. Further, not all animals in the same area will change and some individuals turn only partly white, giving them a piebald appearance. It is certainly not the amount of snow lying on the ground that determines how white they become. The snow cover is too undependable. And on numerous occasions I have so easily picked out a brown stoat in the snow, and in the transpose position a jiggling ermine on a black winter moor.

With a squiggle and a jump the ermine ran through the myrtle and shot up a small birch tree to investigate the branches. It leapt from a lower branch to the ground and began tunnelling through the leaf-choked blaeberry, shovelling and snuffling, all at high speed. Then it was gone, back into the grasses and myrtle where vole runs offered more chance of a meal. Altogether, the stoat had been in my view for barely thirty seconds, yet it had managed to cover all that ground. It was an awesome display of a predator's abilities. There was a last glance of its shining black eyes, twitching ears and ever probing, testing nose, all set on its quicksilver body leaving me with a premonition of doom for some vole.

Dawn was over and the sun's golden orb blinked new light over the top of the trees. Flocks of jackdaws and rooks, and pairs of hoodies radiated out over the land from their communal roost in the pines. A heron lifted from a ditch with a ponderous flight. I had disturbed it while it was gorging itself on the puddocks that lay thick and easy to catch as they spawned in a tiny pool of water. On top of a birch, growing tight against a plantation fence, was a kestrel. Its head was held down as it concentrated on the hunt. Farther on a buzzard flapped out of an oak, owl-like at first with its brown rounded wings. There was still no lift in the air so its wingbeats were deep and laborious. The day was slow in awakening.

I walked up through the dark corridor of a plantation to gain the high ground. As I passed successive corners on the track and a corridor where electricity pylons ran through the wood, I could look out to see how high I had climbed since the previous gap. Otherwise the climb was cold, dark and enclosed. When I eventually reached the top the clouds had at last begun to roll on the first breaths of wind. Pockets of mist still lingered by the burnside, but since clearing the woods I could see high up the hill and right over the glen. I stopped for a look around. There were no eagles about.

In recent years, the eagles occupying that patch of ground had been breeding successfully and I simply wanted to have a wander around early in the season to check if a pair of birds was present, in full adult plumage and showing signs of intention to breed there. My plan was to walk up one glen, check the known eyrie there and come back via another glen to give a good round coverage of the area. I carried on up the burnside, leaving the plantation behind. Far up the valley, a thin line of old pines and birches lined the stream and innumerable siskins and redpolls swarmed through the birches in one massive flock like an infestation of arboreal mice. Over on the far bank, a similarly sized flock of redwings and fieldfares whistled and chaffled as they worked through the thin scrubby patches of juniper in a similar commotion.

Forever tripping and slipping on the steep bank, I made things even more difficult for myself by walking with my head in the air looking about all the time. And I was just about reaching the point of exasperation when I noticed a dark clump in the branches of a pine. It looked like an eagle's eyrie though it was a bit small. I never knew of it before, so up I went into the crown to have a closer look. That was the first time I had climbed a tree that year, and it felt like it, my arms were weak and muscles stiff. No worries, though, as I knew I would grow more supple as the season wore on and the number of trees and cliffs climbed would rapidly increase.

Sure enough, it was an eyrie, a brand new one built from scratch that year. All the branches used in its platform were recently broken off and still white at the ends. Within the rough platform was a well formed cup of larch twigs from the plantation and it was lined with a bed of woodrush and purple moor-grass. These latter two plants are favourite lining materials of golden eagles all over the Highlands as they provide a soft but firm and well insulated bedding for the eggs to lie in. A sprig of scots pine on the rim gave the nest a fresh appearance. It lay on top of the frosted grasses, so the birds must have added it earlier that morning. Not wanting to linger and put the birds off using the nest, I clambered down from the tree as quickly and safely as I could on the cold wet branches. My arms hung limp and aching when I touched down, but I had to put the birds first and speed on up the glen away from the nest, over the ridge and out of the birds' view.

When I was only a few hundred metres farther up the burn I glanced up to the ridge on my right and there I saw an eagle, a full adult eagle, hanging in the air. It was undoubtedly watching me. Whether it was hunting or watching me earlier while I was at the nest I do not know. If it had been the latter, it was not put off as that pair went on to lay eggs in the nest and successfully rear a chick.

As I tramped through the heather on my way up to the ridge a cock red grouse tuk-tukked at me, and I thought how his bones might be adorning that eyrie before the end of the year. I didn't rest until I was well over the skyline and even then only slowed down, preferring to catch my breath as I walked down the far side of the hill, rather than sit down and grow cold. My best route lay right down by the stream and I had just gained an appropriate deer-path when a trio of blackcock burst from a copse of alder and aspen by the burn. They had been hidden on the ground amongst the rushes and birch saplings. I wondered if some of those branches might be used for the eyrie. Aspen is now a scarce tree in the Highlands, and most of their remnant stands cling to

Some eagles are spoiled for choice of nest sites. This one cliff holds no fewer than seventeen eyries, four of which can be seen here. The largest and most often used eyrie is in the tall chimney crack and it is over five metres high, filling the corner so that there is now barely room for the eagles to squeeze onto the nest. It is often the case that within a home range there is one nest that is favoured more than the others. Old eyries fade to grey and are difficult to see against a grey cliff.

 - The big nest in this photograph has since fallen out and eagles have begun building a new series of eyries in the corner.

inaccessible gully walls and cliffs. Yet as eagles frequently use it to decorate their eyries, they must deliberately select it for some reason, and that stand was barely a kilometre as the eagle flies from the eyrie on the other side of the hill.

Four roe then jumped up. They had been feeding on the fresh growth of cotton grass by the water, I could see the freshly nipped off stems. After these sudden surprises I kept my eyes well open and continuously scanned my surroundings. Several mountain hares sat on the slopes above, obvious white dots, keeping motionless, trusting on their stillness to hide them from the eagles' eyes. The mountain hares' colour change from brown to white overwinter is in response to the change in day-length rather than snow cover. Unfortunately, when there is no snow lying the hares are rather obvious, but as those animals were in a group and all alert, they would have seen any eagle approaching. Eagles don't usually chase hares, but rather use surprise attack strategy, so those hares would have been safe as they would have likely out run and out-manoeuvred any eagle. The whole area was rich in wildlife, there was abundant food for the eagles, and on reconsideration I gave better odds on the cock grouse's fate.

For a second, a peregrine flitted over the skyline in a fast low flight, then was gone. A hen-harrier swirled and reeled its wings along the hillside, quartering every inch of the ground, hunting for voles. The sky was still clear, although the day was near done as I made my way back off the hill. Ripples of water had begun to freeze on slabs of pink granite now that sun was off them. There was a hard frost on the way, and I noticed that there was a buzz of bird life on the hillside to my left which the sun was still shining on but there was none on the dark shaded side. The short urgent rush of activity on a fine spring day had passed with the oncome of another winter's night. Hormonal levels would be a little higher the next day. They would further affect the pelage of the stoats and hares as well as their behaviour. And the eagles would pay a little more attention to their eyries as the days lengthened.

It was unfortunate that I had missed the eagles at their eyrie that morning. And while I was hurrying away I visualised the birds' antics. They would have broken off a sprig from a nearby pine, flown across to the eyrie and landed gently in an upward swoop. They would have delicately placed the fresh greenery prominently on the rim. Golden eagles decorate their nests with fresh greenery, be it pine needles, holly, juniper, woodrush or rowan, well through the breeding season, from the initial building to when the chicks are almost fledged. Nor is it only the nest that is in use that is marked, other nests in the birds' home ranges are also adorned with a token. And although most such work is done in spring, I have even seen fresh greenery on eyries in November. Exactly why they do this is unknown, it might be a territorial display, leaving their mark on any eyrie in their home range. Golden eagles are large successful predators and once they have killed and eaten they spend many hours sitting around digesting their food. All that 'spare time' may give them a chance to develop complex behaviour so often displayed by large carnivores. Whatever the reason is, the adding of a single sprig of greenery to a nest does not compare with the full blown effort involved in building an eyrie.

The most industrious construction that I have witnessed was by a pair of eagles a few years previous not too far from the eyrie mentioned above. By parking my car on a high moorland roadside I had strategically placed myself to spy a nest cliff through the telescope before walking in. Although the weather was not terribly bad there were frequent squalls of sleet blowing through the glen on a bitter heat-stealing wind and I was reluctant to go out onto the hill. Even so, the wind-chill was bitingly cold as I spied through the open car window.

The nest cliff was in the lee of a hill and was probably a much warmer place than my spot. I angled the telescope and focused it fine, and to finish the picture an eagle flew straight into the scene with a twist of heather in its talons. It crossed to the right, I swung around a few degrees and watched as it landed on a ledge of the cliff where it proceeded to stamp around and tuck the heather into the thin shape of a nest. What a stroke of luck. I didn't need to go out on the hill that day after all. The eagle was accompanied by its mate and both birds carried on building that nest for more than two hours. How long had they been there before I first saw them?

In perfect order, the male and female took turns to add material to the nest. The female noticeably spent more time and attention on how and where the material was added. The male tended to either simply drop his contributions onto the pile or give them the tiniest of tucks into place. When the female came in behind him she rearranged any of his additions to her own satisfaction, and then added her own piece to just the right place. Most of the building material was heather plucked from the steep slope at the back of the nest corrie, about two hundred metres from the nest and on the same level. Both birds took a straight flight to and fro, occasionally veering up and out on the wind to reach the heather bank, but always flying back to the nest in a straight glide. To pick up the heather they landed on the slope, awkwardly, took off downhill with a firm grip of a strand and pulled it out of the ground by the roots. On one occasion, the male took hold of a strand that stretched for two metres and still never pulled it out. He tried once again, tugging and bouncing up and down in the tall heather, but he failed and flew back up to a some heather higher up the slope where he pulled off a more modest sized strand. The hen broke off a small limb from a fallen pine tree and added that to the eyrie, and both birds added a few tussocks of grass and moss. Branches of pine, rowan or birch look stronger, but they are not so easy to gather nor as supple as heather. Long twists of heather are springy and form a good solid, yet flexible structure, and it take years to rot.

Two hours later, the birds were still at work, but a shower closed the show by obscuring my view. My hands and feet were so frozen after sitting in the wind for so long that I felt I would have been better off walking over the hills after all. I would certainly have been warmer than I was sitting in the full force of snowstorm. The eagles, however, showed very little attention to the weather, something that I have noticed before about these birds when building their eyries. Eagles will build, or at least add material to their nests during showers of rain or snow, although not during torrential rain, prolonged snowfall, or strong winds. Powerful though they are, even eagles cannot carry heavy or long wind-catching branches in the worst of weather.

As spring approaches, golden eagles are driven to prepare for the breeding season by a hormonal response to the lengthening daylight. Weather is a secondary factor in this drive, and a frail obstacle to the urge. When the days are drawing out and there is a drive to build their eyries, eagles are liable to do so at any daylight hour. I have visited an eyrie early one spring morning the day after a fierce snow storm. I expected to have to dig snow off the nest to look for any signs of activity, but the birds had already done so and added a neat ring of branches. The example above took place between nine and eleven-o'-clock in the morning, and I have watched birds busying themselves in the afternoon. Although, with regards to evening activity, I can only remember seeing two episodes of birds bringing in branches. Both were on the west coast and the later light there might have influenced their behaviour.

What materials and how much of them eagles use to build their nests depends on where in the country they are nesting. In some parts of the eastern Highlands where golden eagles nest in scots pines there are some old traditional nest trees which are known to have been used for more than a hundred years. One of these nests, or rather a collection of several nests on top of one another, eventually filled the tree so much that the incubating bird was actually on the crown of the tree while the base of the eyrie was several metres below her. The weight of that nest must have been measured in tons. A large portion of the central column of the nest was blown out and in subsequent years the birds began to nest on the lower section again. Typical in material, if exaggerated in quantity, that eyrie was mostly made of scots pine branches with a little heather and birch. The lining of the nest has always been a succulent green base of soft scots pine sprigs with a soft spread of dried purple moor-grass leaves. The broken ends of the pine branches and the bed of needles give off a fine fresh fragrance – what a wonderful home.

Eyries in the western Highlands or Hebrides can be as large as the nest described but they are mostly set on cliffs and their architecture usually governs the maximum size of nest. Some nests are bulky broad structures tucked under overhangs for shelter, others are tall slim columns filling up corners or cracks, and many are simple rings of heather barely a hand high and laid on a ledge or steep rocky bank.

Many of these nests are blown apart by the westerly gales soon after the birds have fledged their young or otherwise abandoned their breeding attempt having lost their eggs or young. While still in use, as more and more material is added to the nest during incubation and the young grow, the nests can grow considerably. The eggs might be laid in a single ring of heather lined with grass, but the chick might fledge from a thirty centimetre mattress. Such later added material is never properly woven into the eyrie, however, and it is soon blown out by autumn gales.

Heather is the main material in western eyries unless a site is close to some trees. And the lining is often dull compared with those of the pine eyries, consisting of mosses, dried grasses (mostly purple moor-grass) and greater woodrush. Usually there is a touch of colour in a fresh clump of lush green woodrush, a twig of rowan or a sprig of holly. Rhododendron leaves are an unusual decoration that is probably snatched from around one of the gardens of big houses where this rambling weed has been introduced. Other oddities occasionally brighten up eagles' eyries. Red deer antlers are the most common of these. In order of frequency, I have seen the aluminium rods of weather balloons that have crash-landed on the hillsides (these litter all over the Highlands), rams' horns, old fence posts are locally common, and I've even seen a walking stick in one. Whether that stick was left accidentally by a visitor to the nest or picked up from the moors and brought to the eyrie by an eagle are equal possibilities that might explain its appearance.

Within the subsequent layers of the eyries there are often collections of bones. The dried and bleached hind leg bones of mountain hares and rabbits last for years as they are crooked and tie firmly into the matrix of a nest. Those of grouse and other birds are softer, they do not weather so well and are less often found.

Whenever I visit an eagles' eyrie I find it fascinating how their structure and contents reflect the different habitat, plants and animals in the varied Highland terrain.

...s of use, the eagles have built up the lower nest in this tree too high for them...
...y under the branches above. So, they have built another eyrie on the next l...
...e is more room. (Ewan is again the climber)

The usual golden eagle clutch size is two, and often only one or no chicks are reared. However, in areas where food is plentiful, such as around this eyrie in the eastern Highlands, they can lay three eggs in a clutch, and even rear three chicks.

Golden eagles line their eyries with greenery and in Scotland this is often clumps of greater woodrush, sprigs of rowan or birch. In this case in a tree nest in an old fir, they have used surprisingly soft Scots pine needles.

9

INCUBATION

Golden eagles incubate their eggs for about six weeks and during that time the hen, who does most of the incubation, must endure the last of the winter weather before the eggs hatch in the first days of spring.

Sutherland's hills were capped with a late fall of spring snow. The whiteness added extra loft to their posture above the low spreading moorland, and black water deepened the peaty hollows down in the bog. Ragged outcrops of lewisian gneiss burst through the worsted heather and purple moor-grass. There was only a hint of spring's arrival on the brown landscape, where under close examination, green shoots could be seen on the heather and crimson buds were splitting open on the birches.

Grey north-west winds carrying heavy blocks of showers swept in from the Atlantic. They brought cold sharp rain. There was no protection from them out on the bogs. Everything was exposed to the elements' whims. Not much wildlife was out and about on that day, and a single cock red grouse leapt into the wind from the comfort of a heather tussock, scolding me for threatening his security. Down in a wet rush-filled hollow, three red deer stags stood rear-end to the wind, their heads all down tugging at the heather. One of their recently cast antlers lay on a deer path. Chewed well down from the points by the deer themselves, it was a stark sign of the thin living on the peaty moorland. The deer eat cast antlers to supplement their meagre calcium intake from the nutrient-poor heath and grasses found on the moor. One grey-coated stag lifted its head, shook it, then its shoulders, and its hindquarters all in sequence, casting a halo of spray around itself. Now a little drier, it lowered its head and recommenced feeding.

On turning a ridge into a far corrie, I spied over the choppy water of its loch with my binoculars. No black-throated divers were fishing there that day in those wild conditions, which was a pity as I had been looking forward to seeing them. In compensation, over in the far bay, there were four handsome whooper swans. They were riding the waves in style, dipping and stretching their long necks under water to nibble the sparse growth of pond weeds. Above the enclosed side of the loch, a great buttress of rock swept from the snow-line on the hill down to the shore. Sheets of white water were not so much running off, as being blown off, the crag. Several waterfalls tripped over the rocks, and white veils whipped back up over the cheek bone of the face. A mere trickle reached the scree-choked throat of the gully.

The prospect looked bleak as I scanned the cliff from the lee of a rock. There were no obvious signs of birds being present; no fresh droppings, no fresh material added to either of the two most often used nests below the large overhang. The nest site appeared to be deserted.

Jumping from boulder to boulder, I followed the rim of the loch around to the foot of a massive lump of ice smoothed bedrock opposite the nest cliff, then clambered effortlessly up a short gully, lifted by whirlwinds gusting up from the loch. On top of the promontory, the wind was so powerful that I could scarcely stand upright, so I prostrated myself in obedience of the lashing squalls and lay behind the steep side of a roche moutonnée. A scattering of food pellets and cast feathers around me indicated that my sheltering rock was a favoured perch of the birds and that they had been there recently, probably earlier that day.

Spidery waterspouts soaked me as they spun from the loch and leapt up onto the high terraces where they washed the red sandstone screes. I tucked myself into a secure but scanty refuge and dug out my binoculars and telescope from the rucksack. The view through the binoculars suggested that the birds were breeding after all. There was a third eyrie set tight into a right-angled corner on the edge of a narrow straight-sided gully. It had been invisible from below. Once I had set up the telescope and squinted my eye in the draught, all was revealed. Sitting, or rather, lying on the nest was a fine adult golden eagle. The bird's head was held low, away from the wind. Its body was turned oblique to the corner. Water droplet patterns showed that although the eyrie was escaping the full force of the storm by the neatest of cuts, it was still catching a faint eddy. More than an eddy tugged at the telescope, and it misted up continually in the wet air. In no time the front lens was awash with water, wind driven rain. I was struggling.

Meanwhile the eagle sat relaxed on its nest. Golden nape, yellow bill, yellow iris, dark eyebrow, blond back, dark brown primary and tail feathers swept out behind, all of its feathers could be picked out as it quietly incubated its eggs. A stray secondary feather, caught by the wind, twirled out of line. The bird tenderly tucked it back into place with its bill, and before resettling with its bill under a wing, it cast a penetrating glance in my direction, a long thorough look.

One of the most dramatic wildlife images was before me - a lone eagle sitting on its bulky stick eyrie, perched on a tiny rock ledge overhanging a sheer drop to the foot of a cliff. Eagles have been portrayed in various versions of this scene by authors and artists, many inaccurate or fantasy. Here was the reality; there was no need to contrive. The setting was deep amid storm-bound peaks. Bare rock glistened, blasted clean by the rain raging against the cliff. And the eagle sat firm with an air of superiority, as if knowing that it could sit out the fiercest of storms in its impregnable throne set high above the war torn landscape.

I left the bird sitting there as I had found it, and enlightened after my audience, withdrew from the corrie. The strengthening wind forced me to thrust my hands deep and firmly into my pockets to prevent the gusts filling my jacket and lifting me off my feet. The very same wind which I feared, the eagle could ride with command.

Clambering around the hills on one's own can be a bit of a commitment as far as life is concerned. I have never ever been frightened in the hills – apart from when under an adrenaline rush while rock or ice-climbing. That is different, and reasonably safe as one is usually on a rope. I feel confident and in my element when exploring far corries or scrambling along hillsides, and being alone has never worried me. I enjoy it. But what if I had an accident? Well, there have been some close situations. Not so much when climbing into eyries, in those situations attention is held by concentration and care goes naturally. More accidents or near accidents happen in the hills when people are walking than when climbing, and a simple trip or slip can lead to a serious fall. One such incident happened while I was clambering down a steep slope with lots of exposed slabs of slippery quartzite. I had been trying to be safe and kept off the bare rock as much as possible. And one ledge looked perfectly safe so I stepped firmly onto it while with no handholds. Oops! The moss slipped and crumpled up like a carpet on the smooth rock beneath.

For a second I rushed towards the edge and quite honestly, I was helpless. Suddenly, as in all scary stories with a happy ending - I stopped on the very lip of a very long drop. With what seemed like a long hold of my breath, I slowly squatted down and poked my fingers around in the moss searching for a firm grip. My fingers hooked onto a tiny square edge. I turned and thrust my other hand into a crack and pulled up off the ledge. Then I scampered off to the tune of a few extra heartbeats.

That incident happened many years ago now and since then personal safety in the hills has been helped by technology in the form of mobile telephones, GPS navigation and EPURB alarms. However, even when connected to the outside world by these devices, if alone in the hills the primary rule is to take care at all times, especially when on what seems easy ground. For it is obvious to be alert in precarious situations, but easy to be caught off guard once away from it. Gadgets only aid recovery if physical conditions and communications allow, prevention is far better.

It had been raining that day too, and that had probably affected my judgement. Incubating eagles have to tolerate a wide range in weather, and anyone who studies eagles must also live with what comes, take the good with the bad, and adapt.

The weather in eastern Scotland can be so different from that of the west. The eastern hills hold more snow, whereas on the hills of the west it is milder and they receive a higher annual rainfall. Winter generally lasts longer in the corries of the eastern Highlands. Open skies may give sunshine and warmth during the day, however, at night the same open skies bring hard frosts which compact the snow in the high corries into firm, long lasting névé. One year, when I was out on a round checking the welfare of the eagles after a week of such frosts, a spring snowfall settled deep and soft on top of the old névé.

My way up to a nesting corrie lay through a pine wood. The pines stood in soft silence. Outside their cover it was snowing heavily, very few snow flakes penetrated the canopy. Only a whisper of wind found its way through the branches. A scurry of red deer hinds trotted along the top edge of the wood. They kept their distance from me although it was clear that they were not keen to leave the shelter of the trees. Likewise, I followed the shelter of the wood as far as I could before striking on up the open hillside towards a bealach. The hinds stood still in a line, watching me as I ploughed through the snow. When I was well away from them, they relaxed their long necks and tripped back into their refuge. I envied the deers' long, high-stepping legs. Knee-deep snow when lying on top of knee-deep heather measures up to a thigh deep trudge. My progress through this formidable barrier was repeatedly halted for gasps and curses as my feet tangled and tripped in the heather stem snares which sent me flat into the sleeve-filling powdery snow. Picking a route was not easy. Every feature was absorbed into a gently undulating snow-slope. I looked back and laughed at my meandering trail, a dark line of stipples rent by deep tears where I had fallen.

As I reached the bealach, the walking became much easier. The wind blowing through the gap had prevented any deep accumulations of snow. Down in the corrie, on the far side, the snow became deeper again, and what had began as a light flurry of snow in the morning had now become a full blown storm. Then, as April showers will do, the clouds rolled by and the snow stopped falling. The rest of the day was left to the blue skies of spring.

A shallow gully led down to the top of the cliffs which I was out to check. The snow was deep, but there was less heather beneath so walking was easier. I shuffled around the top lip of the crags. They were set in two distinct sections divided by a steep open slope of heather and woodrush which was best avoided in those slippery conditions. Slowly and quietly, I tried to find the top of the western buttress where I knew of two established eyries. By creeping onto the snow-filled ledges on dubious footholds I could, at a pinch, look along the face to the first nest. A deep pillow of snow on top of it told me that it had not been touched by an eagle all year. The other eyrie which I knew of was tucked into a chimney, a vertical gap in the cliff, a few metres further along. There was only a shallow topping of fresh snow on the second nest, so I dropped down to the nest ledge for a closer inspection. Fresh sprigs of rowan and pine and a tuft of woodrush indicated that there was indeed a pair of eagles occupying the territory, although they had not chosen to lay their eggs in that nest either. The shallow ring of nesting material was possibly a result of the birds' nesting display, or they had changed this original choice of nest for another.

I climbed back up to the top of the cliff, took a seat in the snow and looked around for a sign to lead me onto a third eyrie which I did not know. The sky was empty. There were no eagles up. Down in the glen, black dots of deer spread out from the wood towards a wind cleared ridge where they grazed hungrily on the heather. Below the eastern buttress, a wren gave a short burst of song, and ticked away to itself as it probed the dark hollows beneath a tumble of snow-capped boulders. I gave it a spy through the binoculars, and was just wondering how such a frail bird could live in such a harsh landscape, when I glimpsed something on the cliff up above. There it was. A jumble of branches poking out from a nook in a cliff could only be the nest of an eagle. Only the edge of the nest was visible, the cup was out of view. With a new confidence, I jumped to my feet and was on my way over there.

The eastern buttress was more difficult to view than the other. At first, I could not gain a vantage point from which to see if the nest was in use or not. Unsure of the lay-out, I carefully traced around the cliff-top. It was high, well above the level of the eyrie and a thick growth of rowan trees obscured any view down the cliff. My eyes were peering down, looking for a round nest cup, possibly with eggs. I had expected any sitting bird to have slipped off the nest while I was busy across on the other buttress. Obviously not so, ten metres down a v-shaped groove I could see an eagle sitting tight on its eyrie. The western cliff was out of its view. The bird was a female. She was large, with her fine dark adult tail laid out over the rim of nest behind her. All around the rim of the nest there was a thick ring of fresh snow emphasising her broad head and shoulders. She must have been sitting on her eyrie all through the snow storm earlier in the morning. I kept absolutely still, watching her through a lattice of rowan and heather. She never knew I was there. Her head turned back and forth from left to right as she watched a troop of crows straggling through the glen. She never looked up. I had found what I wanted, there was no need to disturb her, so I backed away from the edge with the knowledge that she was safe and secure.

Now that I had dropped concentration I noticed that the frost was nipping at my toes. They soon warmed up as I climbed up out of the blue shadow and into the white sunshine. That bird had chosen a well sheltered site for its nest. Set out of the wind, relatively little snow had fallen on the-eyrie. Cold as it looked; the sitting eagle was well insulated by thick layers of feathers and down. Eagles have been known to sit through greater snow storms than that one. Just how bad a storm they can sit through I do not know. I have seen one nest deserted because of a storm while its neighbour was not. And one year, four nests in adjacent territories were deserted after a severe blizzard. Those four nests usually escape the prevailing westerly winds because they face east. However, that blizzard was fierce, long-lasting, and it blew from the east. Adverse weather was undoubtedly the cause of the failure. Golden eagles breed in the lower arctic, so it is not only the cold which causes them to fail. The severity and length of exposure to storms are critical. And prolonged rain as well as snow can cause a desertion as I have seen more than once in the western highlands after a torrential battering of rain.

Eagles might give up their nesting attempts due to direct exposure to the weather or they might do so through hunger, for if the weather is rough, they would have difficulty hunting. A lot can happen in six weeks.

Only a few days old and just able to hold its head up. This chick has a healthy look about it - bright eyes, a full crop, chirpy, and investigating everything lying about the nest. The parents of this bird brought numerous hedgehogs to the nest, and delicately fed them to the youngster. As often happens, the second egg failed to hatch.

Purple saxifrage is one of the earliest of the alpine flowers to bloom, coinciding with the eagles laying their eggs. Bright clumps hang from wet seeping cracks in the rock below a number of eagle eyries, adding flecks of colour to the cold grey cliffs.

10

MOUNTAIN PLANTS

Eagle eyries are often surrounded by wonderful hanging gardens of alpine plants.
Some are rare, most are common, all add colour and character.

High on the flank of a north-west Highland hill there is a great wall of rock that
gives shelter to a set of eyries; each tucked away from the westerly storms. In
early springtime the hillside in the glen below is a rolling sea of dull brown
heather and dead withered grass of the previous year's growth. On the cliff,
black streaks of slime slither down, green mosses abound, and slippery green
leaves of woodrush shine as the wind ripples through them. The cliff is remote,
and wild in the sense that the vegetation is in a near-natural state, unaltered by
man apart from airborne pollution.

By summer the scene is no less remote, although the cliffs lose some of their
austerity. The rock is drier and safer to climb, the warm sun makes for a casual
walk-in, and there is colour in the hill. The heather has a flush of fresh green in
its shoots and the new grass stems are well on their way to full length. Along the
path-side, there are specks and flecks of yellow tormentil, blue milkwort and
pink lousewort. And all these would be considered a fine display if there was not
a finer one nearby.

Up by the eyries there is a far grander show. A long hanging meadow lies
at an angle of about sixty degrees below the main nest cliff, where a thick mat
of vegetation overlies a treacherous slab. It is wet and edges have frayed off in

a brown ooze from the underlay of mosses, and that is all that seems to hold the whole field to the slabs. I first saw that meadow in bloom when the tall stems of yellow globeflower were shining, backlit by the sun. Drooping hepatic heads of water-avens bowed to the crowded heads of angelica. Sprays of red campion, the strongest colour, dominated the sward. And in the wet cracks on the walls, fleshy lobes of roseroot twisted their roots for a hold, their yellow flowers freckling the grey.

A few feet above the top edge of the meadow, the white downy head of an eagle chick peeped out on the world below, a wide bleak world seen through a roseroot tinted foreground. At the back of the nest ledge a rowan has taken route. Its limbs are thin and the bark is grey and peeling. It might be a young sapling, but more likely it is a very old tree indeed, merely stunted in its growth by the physical restrictions of the cleft where the seed was once dropped by a thrush. A thrush, a redwing, fieldfare or perhaps a blackbird, which had probably flown over on migration from Scandinavia, would have eaten a few berries in the first woods it found upon landfall in Scotland. Then it would have passed seeds in its droppings as it foraged on the richly vegetated cliffs, before flying farther south and west as the winter came on. And from one of those seeds, the rowan tree grew.

Some of the rare plants that grow on the eagle cliffs only grow in montane situations, and give a special enjoyment when one comes across them. The more common mountain ledge plants are not confined to the cliffs. Many are species that simply have become less common below as man has cleared woodland and his beasts have grazed out the more luxuriant herbs from the ground cover of heaths and grasses. The same cliffs that harbour eagles also give refuge to plants.

When climbing into eyries, it's all very well grabbing tufts of woodrush or heather for handholds, although care should be taken as the roots of many of these cliff-dwelling plants are often set in thin soil. They could pull out at any time. So I tend to hold rather than pull, and tread lightly, testing the ground before committing my full weight. I much prefer to climb on the rock, but that is not always possible. And that is when knowledge of plants comes in useful, preventing damage to any rare plants. I was once up at an eyrie near which there are several ledges full of the scarce yellow oxytropis. It took a bit of tactful route finding and delicate footwork, and with a bit of care, it was reasonably safe - for both plant and person - to dodge and so preserve the plants at the risk of my own life. Balance is always important in climbing.

By the time I clamber around on the final checks of the eyries, harebells have usually spread out in a blue wash on the drier ledges, mixed with spikes of golden rod and hawkweed. Small tortoiseshell butterflies are common amongst these flower beds and the whole has an air of a quiet cottage garden. Except some of the neighbouring ledges hold alpine specialities like the tiny

Foxgloves frequently grow on eagle nest ledges, whether in use or unused. And they persist on ledges long after the last remains of an eyrie have decayed.

Little drooping petals of dog violets catch the slanting sunlight.

This is a common flower below long-used eyrie sites, where the birds' droppings and prey re-mains have rotted into and enriched the soil. The commonly abundant heather of Scottish hillsides is less dominant in these places and small patches of herb-rich grassy heath have developed. Eagles have probably used the same ledges for their nest and roost sites for thousands of years, and the herbs gardens have developed as the soil became enriched. And I know of several less rich herb patches which have become established beneath tree eyries, indicating that these too have probably been used - for over a hundred years in at least one case. Other flowers which thrive on these fertile grounds include wood anemone, tormentil, bird's-foot-trefoil, heath milkwort, primrose, heath bedstraw, foxglove, ribwort plantain, devil's-bit scabious, harebell and marsh thistle.

Eagles live in the mountains, and mountains are made of rock; with crystals, strata, crack lines
and lichens. Here the rusty-red colouration of one species of lichen is due to mineralisation in
the rock. There are at least five species of crustose lichens on this one piece of rock on a hillside
in Perthshire, and they would require microscopic and chemical analyses, which I did not have
at the time, to determine their species,. There are hundreds of species in the Highlands, growing
in a multitude of coloured patterns and textures.

bells of alpine gentian, or hanging sprays of rock speedwell with its blue, red, and white petal targets.

Perhaps the most exuberant cottage plant that grows around eagle nests is the foxglove and tall ranks sway around in the cliffs' breezes. This seems such a large plant to be growing so high on a cliff, yet it is not unique, the even taller and rare alpine sow thistle also grows on remote ledges, although not next to any eagle eyries I know. Those luxuriant ledges are some of these plants' last stands. And these plants share one same plight as the eagles. The easier accessible ledges are attracting more and more visitors. One ledge I know now has a well worn path leading to it and onto it. The plant is being trampled. There is no need to do this, I have never stood on the ledge, preferring to simply spy it from below through binoculars, although even these are unnecessary as the plant is so large.

Another plant that grows tall on eagle cliffs is the common rosebay willowherb, the same plant that is so abundant on wasted ground around our cities. I know of one eyrie where there is a field of it draped down the slope below. Golden eagles add sprigs of greenery to their nests, and the willowherb could have been introduced that way, then spread as it cast its clouds of airborne seeds. Although the reason for eagles adding foliage to their eyries is speculative, one purpose could be to freshen the nest as it becomes more and more soiled and fly-ridden as the chicks grow. That is certainly how I appreciate it.

The plants most commonly used by eagles as 'air-freshener' or 'sanitation' are greater woodrush, rowan, birch, Scots pine, holly and juniper. One species which I have always expected might be used is honeysuckle, yet I have never encountered a sprig of it on an eyrie. I know of several eyries that have sprays growing over them and one vigorous plant has well and truly entwined itself through the grey sticks of a nest. The scent there must be so heavy on a warm summer's night – and what moths are attracted to those distant cliffs? Then there are some eyries, mostly in the west, which have been crept over by ivy. The whole wall of one cliff, about twenty metres high, is covered, and there are three eyries hidden by its coils. Wild rose is another climber that sometimes grows by eyries, and perhaps it is because it is so vigorous about the nest ledges that the birds have not used those for a long time.

Both golden and white-tailed eagles nest on sea cliffs in the west, and some alpine plants also grow there. Thrift, roseroot and purple saxifrage are three species, which might be considered the more common alpines that also thrive in the maritime zone. Then there are the many common species, some of which might be considered as hedgerow species, that all occur on the high ground also. These include angelica, hogweed, bluebells, hazel, and aspen. And there is a host of common grassland and heathland plants, which are found on both sea and mountain cliffs.

The patches of ground below eagle eyries should perhaps be considered as a separate specialised habitat niche for plants. For the sward in such situations usually takes the form of rich green growth, which shows up brighter than the surrounding heath. And molehills are often found there, indicating the richness of the worm-bearing loam. There is one vibrant example below a nest cliff in Argyll, which has a southerly aspect above a steep slope with a well-drained soil. Although the cliff is of basalt, and the resulting soil is already base-rich, the slopes below the eyries and roost sites are particularly rich in flora. And out of that soil grows a tightly cropped sward of sheep fescue, wavy hair-grass and common bent, with daisies, dog violets, harebells, birds' foot trefoil, clover, self-heal, heath speedwell, and devils'-bit scabious. Moonwort, that peculiar fern that looks like a diminutive flowering plant, hides between the grass blades. While up on the cliff itself, there is a rich and thick growth of wild roses, tall stems of sow thistle, and spikes of golden rod. Thyme hangs over the edges and thick tufts of mosses cling to the wetter walls.

Similar enriched niches also occur around grassy knolls where the birds often sit while watching over the land below. The herb-rich grassland patches attract the sheep and deer, and rabbits and hares where they occur. These in turn further enrich the soil with their droppings, although they eat out the herbs and leave only the grasses. It might be that it is the grazers which create the fertile soils on such hillocks, but as there are such sites high on impregnable ledges, it would seem likely that it is the eagles' droppings and castings that initiate the verdant growth.

It has taken ten thousand years, since the last ice retreated from the glens, to develop the Highland soils. And in areas of more nutrient-poor soils, such as in the granite hills of the east, or in the peaty wet heaths in the north-west, the effect of the eagles' fertilisation can be spectacular, creating lush green oases amidst extensive brown ground.

Adult golden eagles take two years to moult and replace a full set of flight feathers as they need to retain as many as possible at a time for maximum manoeuvrability. So, when the females are incubating, they take advantage of so much non-flying time to moult some primaries and usually tail feathers too. They always cast their feathers in symmetrical pairs from each wing, to maintain their balance. The patterns of speckling on these feathers are as individual as our fingerprints. The same pattern can be seen in the feathers below an eyrie over many years, and then when a change is noticed, it is likely that a new female has taken over that nest

11

CHANGE OVER

There is one eagle's nest cliff that I shall always particularly enjoy visiting. It is possible to view the contents of any of the alternative nests by spying the site with a telescope from across the glen. And one advantage of checking nests from a distance is the opportunity of observing eagles behaving in a natural and undisturbed way.

The eyries are set in a tremendous situation. A great rambling buttress, cut off from the main cliff by a deep wet gully, abuts a vertical grey wall. In the angle where the two meet, a pillar of rock is topped by one nest. Half way up the pillar a square-cut ledge offers a site for a second. And a third nest, the most frequently used and largest at about two metres deep, lies at the foot of the pillar behind a sturdy rowan tree. Below the main nest is a steep, wet vegetated cliff exposed to northerly gales, but all the nests are safely tucked into corners. Close by, under an overhang at the base of the grey wall, a step of rock is white with droppings and cast feathers dropped by the eagles when they take roost. Any water from the cliff above drops well clear of the nests and roost. The birds have chosen their sites well and there are even more alternative nests and roosts established farther along the cliff if the weather conditions should exclude the use of these three.

The nest cliff straddles a high pass through the hills. And below the cliff there is a loch. Storms and blizzards charge though the gap from one glen to another, driving their rain and snow over the broken peat banks and heather.

Even the warmest of days attract turbulent air through the pass, which is well used by birds. Common gulls regularly flit through the gap, spiralling up from the glens on the channelling wind. And red-throated divers frequent the loch each summer. They rest on the wide water while their mates incubate their eggs on the shores of small lochans on the moors up above the eagle eyries, or, they fish for small trout to bring up to their young on the fish-less lochans. In the evening, the divers' songs echo though the gap in the mischievous air.

Greenshank feed along the shore of the loch, especially where it spills out, probing deep between the round lichen-clad boulders. A burn dives down from the loch into the northern glen in a sequence of waterfalls and pools, scouring the rock white. Dippers build their nest under the arched spray of one particularly spectacular fall. And ravens and hoodies build their nests on the scattering of cliffs that spread down the flanks of the glen.

A broad shouldered ridge divides this glen from its northern neighbour where a second pair of eagles nest. To the east, there is another home range, and two more pairs usually occupy the ground to the west and south. No fewer than four neighbouring eagle home ranges' nest cliffs are within view of an eagle sitting on top of the eagle's cliff in the gap. And I have watched eagles perched there on the very summit, looking all around – just how far can they see? There are even more nest cliffs out of sight, set deep in corries within the horizon of high hills.

By going into the hills via one glen and walking over the moorland at its head, transversely crossing the pass, climbing over the top behind the big nest cliff to drop into another glen, then again climbing out over a ridge and down, and finally leaving the hills via another long glen, it is possible to visit at least three of these eagle sites on one trip. Occasionally, four sites can be checked on this route, if another pair obligingly nest in the top corrie in their home range. On about my third trip there, I worked out the most efficient route, the best spots to spy from, and the easiest walking between them.

The first home range visited on this route has a selection of impregnable nests set mid height, on horribly vegetated and crumbling cliffs, of humbling dimensions. One old grey nest can, with contrivance, be reached if necessary, but the others are best spied with the telescope. So, in order to see if a bird was sitting on any of the nests, I had to climb high up the opposite side of the glen, well off the easy walking by the burn side. That glen is a favourite with stags, and there is usually a group of forty there throughout the summer. The most trouble-free way through the leg-tugging heather is to follow their well-worn trails. Ever near, but not quite high enough, I repeatedly stopped and spied across the glen. Then at last, I was able to see an eagle's head staring at me out over the rim of an eyrie. The home range was occupied, eggs had been laid and they were being incubated. All was well.

I had climbed enough, so I contoured round to the head of the glen where I scrambled about for prey remains and cast pellets below a roost under a small crag. The same sheltered crag has attracted eagles for generations. There is an old nest there which I have never seen used. The birds mainly use the cliff as a favourite roosting place. That day there were three pellets below the cliff. One contained hair from the remains of a red deer which would have been eaten as carrion, and the others were mostly composed of white mountain hare fur.

As I tread on over the seemingly endless wet moorland, time and effort were meticulously measured. Then as I neared the far edge of a plateau I hurried my stride, picking a way down the steep broken side of a burn, and settled on the dry sunny south-facing hillside. I lay with my head propped on my rucksack, out of the cold wind that retarded all life up on the moorland behind me. The panorama from that point on that April day was particularly clear, and I traced out a line linking the landmarks before me on the map in my lap. In the southern glen the birches were newly opened into a shimmering silky green. And closer at hand, on the warm hillside where the wind didn't bite, a tormentil's four petals shone bright.

Methodically, I spied each nest on the cliff opposite in turn. Then, having picked out the head and tail of a bird lying low in the second pillar nest. I remembered how I once took a rest at that very same spot to break the long day's trek. With one eye on the nest through the telescope, and the other roaming the hills, I chewed on an apple as I lay catching my breath. The sitting bird cast a few glances in my direction, but on the whole, she ignored me. She never stopped watching the goings-on down in the glen. Every gull that trouped past the nest cliff was vetted with a customs man's eye. A pair of ravens attracted a considerable amount of attention as they tumbled and chased each other about the face of the cliff. And a white downy feather lying loose on the edge of the eyrie persistently escaped the outstretching neck and bill as it twirled in the wind.

The eagle's head tilted up to its left. I followed its line with my spare eye to see a thin strip of wing turn over the top of the nest cliff. It swung down to the level of the eyrie then along past in front of it to land on a grassy ramp on the buttress. The incoming eagle held something in its talons, although I could not see exactly what. I switched the telescope onto it. Golden eagles are exceptionally well camouflaged when sitting on a cliff, and viewed from above they aren't revealed by their distinctive outline as when seen against the sky. The dark wings, blond nape and brown back all faded into the mottled and shadowed background. The yellow talons shuffling awkwardly with their purchase were all that gave that eagle's position away. It then anxiously leapt off the grass to swoop over with a shallow dip and landed on the edge of the nest.

Eagles never drop down from the heavens onto their eyries on outspread wings as so often portrayed in dramatic illustrations. They approach gently from below the level of the nest, and then swing up with a soft flap of their wings which are immediately folded as they stall their flight and touch down. Every eyrie is equipped with a purpose built landing ramp. This is an enlarged straggling edge of the nest platform facing the direction from which the birds normally approach the nest. Once settled on this ramp, the eagles slowly walk onto the rim of the nest cup.

The bird that I was watching dropped its cargo at the face of the sitting bird with an accompanying sequence of several bows. Presently, the sitting bid stood up above the nest cup and tucked whatever was brought by the second bird into the lining of the nest. I took the item to be a clump of woodrush for it was definitely green and straggling. Greater woodrush is a favourite nest lining material of golden eagles and it grows plentifully on wet earthy mountain ledges. Without more ado, the incubating bird stepped gingerly to the edge of the eyrie, dropped onto the updraft and swung over to land on a ledge on the cliff not more than fifty metres away. Immediately, the incoming bird straddled the nest cup, turned around to face out from the cliff and settled down on the eggs. With a jiggle of one wing then the other and a twist of its tail, the great bird was comfortably sitting on the nest. And by stretching its neck over the rim it watched the doings of its mate for a second or so until it landed on its ledge. Then it fidgeted into a low set position, it repeatedly re-aligned some lining in the nest, it turned its whole body into another direction, it re-set its wings, tail and feathers, till finally it dropped its head onto the lip of the cup. The change over of incubating shifts was complete.

For the next ten or twelve minutes the relieved bird re-arranged its plumage. All the flight feathers were re-set to perfection, including those of the breast and belly which had to be resettled after brooding. Then the bird lent forward, lifted its tail and discharged a long white splash out over the cliff. And just to make sure, it repeated the sequence with the feathers. When it had at last completed its toilet, the eagle casually took off and flew silently along the cliff away from the nest, climbed up on the wind and over the hill, presumably to go hunting. It left in the same direction as its mate had arrived.

The second bird was more dozy than the first. Initially it reshuffled the eggs again and again, but soon it lay still. Eagles tend to lie on their nests rather than sit, which, although it is the generally used term for a bird on eggs, implies a more upright position. Its head and bill rested on the edge of the nest and its eyes were closed. It was probably asleep. That day, I was tempted to lie and doze in the sunshine myself, but I had another site to check and a long way to go. So I dropped into the pass, that was easy. Climbing up out the other side around the side of the nest cliff was not. I don't know whether the sitting eagle was watching me as I crossed the glen, although it almost certainly would. Even if it was, I was soon out of its vision deep under the fortress walls and on my way without disturbing it.

Depending on which nest is occupied, it is not always necessary to descend into the next glen to check on the eagles as a fine vantage can be obtained from the dividing ridge between the two home ranges. And that time I was in luck. I focused the telescope on the top nest, and there was a bird sitting on it. All was well. I had spied the first two nests from about a kilometre across their corries, the third I spied from about two kilometres.

My mission was accomplished and from there on the last high ridge, with the satisfying knowledge I now had, the walk out in the evening to the mouth of the final glen was a delightful stroll. And my tired legs, which had pumped hard all day, lightened when they gained the well-laid stalker's path.

As I reached the roadside in the last remains of the day, a song thrush sang from the birch-clad cliffs. The low rays of the sun warmed the fluffed up pink plumage of a kestrel as it settled in a crack. Shadows ran up the cliff past it, and as a family of ravens slipped into a niche in the rock, behind me, the eagles would be beginning the night shift

.

A young eaglet looks tiny in such a large eyrie, but in ten weeks the nest will seem small.

12

NEW LIFE

When eagles lay their eggs in the back end of winter, few birds can be heard in the glens. Then slowly, one by one, while the eagles are incubating, different songs ring through the hills.

The main resident Highland songsters are the cock red grouse that tukk-tukk in the heathered hillsides and the scattering of wrens that blast out their disproportionately loud trills from patches of dark undergrowth. At times there seems to be nothing else except the raw caw of the hoodie crow and the prukk of the raven.

The month of April brings the first body-warming days of the year to the hills. The days become bright and freshening, well on the way to attaining their full summer glory. That is when transient golden plover and meadow pipits move in, visiting the high ground temporarily and retreating to the valleys as warmer weather comes and goes. The plovers' song is the signature tune of the moors and although the pipits' song is so common and seldom listened to, it is heartening to hear their disjointed choruses none the less. One of the finest singers is at his best around the time when the eagles' eggs are freshly laid. He can catch out the unwary, as when I heard a loud whistle from high on the edge of a corrie, clear and full as if from a shepherd to his dog. It was the ring ouzel, up on a lofty rock pulpit, his notes floating far over the glen.

A four week old golden eagle chick squats in its eyrie.

Once they have gained a few weeks' weight the adults leave them alone for much of the day. There are two freshly plucked red grouse on the edge of the nest, but as the chick's crop is full its parents would not come back to feed it for a while.

The chick is by this age filling the nest cup, which is beginning to deteriorate into a broad flat disc, partly via wear and tear of such large birds stepping all over it, and partly by the adults continuing to add fresh branches to the nest during the nestling stage. Why they do this is unknown, but it does keep the nest fresh, and perhaps helps limit infestation of parasites and disease. These birds have brought in mostly Scots pine, with one sprig of birch.

The grouse remains are lying on a distinct doorstep – a sloping edge to the nest where the birds land after approaching from their favoured direction. All eyries have this feature and the branches there are usually spread to form wide, solid landing and take-off platforms.

Some of the finest days in the hills then come in May. Settled in between the immature temperament of April and the indulgent maturity of the full summer, this is when the hills of Scotland enjoy the lively characteristics of youth. It is recollections of May days spent in some of the wilder Highland glens that help to rekindle my own enthusiasm at the beginning of every year. Walking across miles of open bog can be wearisome. Wet sticky peat pulls off boots. Soft sphagnum moss leg-traps tired walkers, countless burns need to be jumped, and the routes relentlessly climb up and down. That is when something as simple as a lilting tui-tui rising in a glen will snap me out of a winter's trudge and into a light-footed exploration of the hills. The song of the greenshank, instantly recognisable.

As the season creeps on, other calls fly in overnight. Down by the river banks where there are shingle and pebbles on the meandering beds, the first of the oystercatchers pipe up and down the glens, at nightime to begin with, then all day. Their loud calls seem to be amplified by the darkness. Later, the river is further enhanced by the repetitive trilling of the common sandpipers as they chase one another from boulder to boulder, rapidly wagging wings and bobbing tails in their restless performance. And grey wagtails keep their inconspicuous notes hidden in the cool rocky gorges where they conceal their nests. Out on the drier hillsides, wheatears scritch and chat as they flick and bob from hole to hole in search of a secret nest site under a rock in the screes. The meadow pipits finally commit themselves to laying their eggs in May and knowing this, the first cuckoos' calls ring over the moors.

The bird song increases as the plants green up the hillsides. Rowan blossom fills windless gullies with a nose-twitching perfume. The weight of the fragrant new growth bows the branches down from their bare erect winter stance. And yellow dots of lesser celandine and wild primrose shine from the wet litter below. Willow warblers arrive precisely when the first trees open their leaves and soon no single piece of scrub that is large enough to house a pair of willow warblers is left untenanted. Chiffchaffs endlessly repeat their double noted song, and ubiquitous chaffinches confuse everything with their urgent twitter and chatter.

I can remember being in one far corner, high out at the head of a glen, when a single male pied flycatcher forced his tiny song above the roaring and splashing of a burn falling into a loch.. He was in a thin line of rowans growing among boulders on the burns banks. That burn runs steeply down a groove cut into the quartzite bedrock. It only runs for about a hundred metres, above that it drops in a graceful white plume from the top of a vertical cliff. The rowans cling to the edge of the water course delicately dodging the jaws of deer and the force of the water. The outside ends of their branches are nibbled and stunted. White washed dead trunks lie straggled in the water, jammed in the rocks where

A seven week old chick lies on its eyrie.
Eaglets spend much of the day just lying in the nest, digesting their food and growing.
Often they will rest their head on the edge of the nest, like this one, watching the world go by below.

spates have dumped them. That bird was so special, being so far from the nearest wood, his usual habitat.

Most bird song is heard below the eagles' eyries, but there are more up above. Snow buntings sing their own rare song from amongst the boulder fields on the highest ground where the snows lie longest. And above the buntings are the displaying female dotterel – their sex roles being reversed - flying around singing above the plateaux to attract mates, especially at dusk. There is a magic moment on the tops after the sun has gone below the horizon, and mist fills the hollows. Dotterel, golden plover and dunlin all sing in the air at once and cast a haunting aura over the plateau. By the time the first dotterel eggs are laid the first eagle eggs have hatched and the frail downy eaglets lie weak in the nest after the effort of hatching.

The season of spring song always seems so short, yet, for the birds, it serves as ample time for attracting a mate and holding a territory in which to rear a brood. The migrant birds are there to feed and rear their young on the abundant insect life of the Scottish Highlands. The eagles incubate for so long, that since laying, the hills have been white, black and then green. They must see so much come and go in that time, and they will have sat through cold days and nights, deafening storms, echoing whistles, the full spring concerto. Soon there will be the tiresome high summer days with no wind. And insects will become a bane to both me and the eagle chicks. Clouds of midges bite, sneaky clegs bite, aggressive mosquitoes bite and creepy ticks bite. The migrants are there to feed on the hordes of insects. If only they could eat a few more.

The best time to see otters is early or late in the day, when the wind is still and the water is flat. If any otters swim by, they inevitably cast distinctive series of chevrons in their wake. I always cast my eye over any smooth water, just in case ......

13

SHORT CUTS AND OTTERS

Part of the joy of looking for eagles is exploring and wandering away from paths into areas where few people ever go.

Many golden eagles in Scotland select the more remote corries and glens to nest in. Some of these are right at the head of the glens enforcing any visitor to endure a long walk-in along river or lochsides before even beginning to climb up to the nest cliff. Among these, there are several eyries which I have initially checked via such long hauls, but subsequently I have gone to them by way of short cuts.

In most cases the 'short cut' has been over untracked ground in contrast to most conventional routes which are usually partly tracked for the use of shepherds, stalkers or foresters. The walking tends to be rough and steep, any cut being merely shorter in distance as the eagle flies. On the ground, saving time usually depends on speed in climbing directly up over a ridge dividing the nesting glen from an adjacent one that holds a public road. Through experience, by trial and error, which includes some terrible time-wasting diversions, I can now check some eagles' eyries more quickly by taking a short rough route rather than the long tracked one. Close attention to landform details help to detect small deviations through gaps, over colls or around either side of bluffs which can all save a little time and energy. And by cannily choosing crossing points over burns, especially when they are in spate, a little time can be chipped off every journey. Although exhausting, while slogging through knee-high heather, every foot placement is made with consideration of where it fits in the big day. A whole lot of energy can be saved by stepping carefully and if less tired, more is seen and heard and that in turn gives a much more satisfying day.

Of course, some routes require more detail than others and I can recall one precisely because of something that happened on the way. I was climbing back out of a nesting corrie having checked a site thoroughly, and finally satisfied that the eagles which hadn't laid eggs in any of their nests known to me had not surreptitiously produced chicks in an unknown eyrie. After a tiring climb up and down and around various buttresses, I made my way out for the last time that year. As I walked out over the moor leading onto the back of a ridge I was reminded how wet the spring had been. The bogs were all full. The peat hags were flooded. Their walls were soft and slimy, and difficult to cross without jumping. I zigzagged from hillock to hillock edging around the lochans. Routinely, I stopped and spied across the waters for any signs of life. In the morning, on my way in, I had heard a red-throated diver flying over the area, calling its weird croaking rattle, and there was a chance it was nesting on one of the truly quiet lochans which lie up on that moor. Unfortunately there were no divers, but there was something down among the reedy shallows at the far end of the largest lochan. And its bobbing and dipping soon revealed it to be an otter, so I lay down to watch it from a relatively dry hummock on the shore, near a small inpouring stream.

With a casual air, as if without a care in the world, the otter swam on towards me. It was concentrating on fishing, so it never saw me. My scent was safe as no wind was blowing and I lay a good two metres above the water level. A gathering of midges did test my endurance however. Every niche of the lochan rim was searched meticulously by the otter. It nosed between rocks lying in the shallows. It swam around the middle of the water. It came onto the shore where it scurried about. Then it was in and out of the water again and again all the way along until it was only a few metres away - directly below me. My body fixed rigid. The midges kept biting. I followed it with my eyes. And midges nipped at my eyes. I peeped through the heather stalks, holding back from lifting up my head for an even better view, then it reached the outflow of the burn behind me. The otter turned in and puddled on upstream. Suddenly it stopped with a jerk and sniffed the light air.

That was at the precise spot where I had crossed the burn not long before. The otter gave another sniff, then a stifled squeak to itself before shuffling away from the suspiciously smelling rocks. It came out of the burn and headed back to the lochan in a straight line. It was obviously familiar with its surroundings for it could not possibly have been able to see the lochan from where it was, yet it appeared to know where it was going. I lay there frozen. And breathless. While totally unconcerned, the otter was completing a ring of less than five metres radius around me. I didn't dare to turn my head to watch it until I heard the footsteps down on the pebbles below. I eased my head around in a long slow movement, expecting to see the otter speeding away in alarm, so sure was I that it would have detected me. Perhaps it had, and assessed me as no threat. Whichever the case may have been, the otter which was a smallish individual and might have been an inexperienced youngster, simply carried on its systematic forage around the shore of the lochan from where it had left off. Then the midges won their battle and I was the one that was off at high speed.

That otter made my laborious hard day into an unforgettable one. And in so doing, engraved everything else that I learned that day into my mind, including the route which I had taken with all its details. It had certainly been well worthwhile taking the less obvious or well trodden route, emphasising the attention and observance which tends to be dropped when following a well laid tack with head down and heavy boots marching along.

Otters are a very good example of the elusive wildlife that can be seen away from well-worn tracks and all their associated human interventions. This is mainly true only for the mainland, the more so the farther south and east. Otters are present in these lowland areas too, it's just that they are especially secretive there and mainly nocturnal. In the far north and west or out on the islands, the otters become more diurnal and exhibiting. In the Outer Hebrides they can be watched fishing, from a car parked on the verge of a road, and in Shetland they can be so confiding that they have been referred to as tame, although this is a widely misused term for animals tolerant of man.

Most of my own personal best experiences with otters have been associated with salt water, either on the shores of Shetland or in the sea lochs of the Hebrides and west coast. On the mainland I have only had a handful of good views. One was while looking out from an eagle's eyrie over a boggy moor. I was sitting in the nest cup, as there was no room anywhere else to sit, while I rummaged through the nest for prey remains. The chick had fledged, so I had no need to rush away. I was admiring the view from the nest, thinking of all the life that the eaglet must have watched while growing up there.

The otter was deep brown and the peaty moor was a very similar colour, my noticing it showed how it is movement that is the biggest betrayal of any animal's presence. My eye was caught by a slight motion. I didn't know what, only that something was creeping around down there on the moor. I focused on the spot, and re-focused until eventually I could make out a dark slithery form twisting through the hags. It proved to be a huge dog otter, perhaps the biggest I have seen, and I have no idea what it was doing there in those hags well away from open water. Otters do wander over dry, or rather less wet moorland quite frequently and there are plenty signs that indicate this, their wide-spread five-toed footprints and their sweet-smelling scats. That one was obviously out for such a ramble, a very meandering ramble, for it was investigating all sorts of holes and ditches, turning this way and that, and double-checking odd smells. Otters move so quickly, they must be able to cover miles in a day, and that one was certainly not taking a short-cut, wherever it was heading.

Several other sightings have been of animals crossing ahead of me while driving, both at night and day, or while sitting in waterside lay-bys. I have also surprised a few otters when watching along streams or peering under bridges for dippers, although none of these sightings were as enthralling as the instances described above. So many wildlife encounters depend on luck, although that in turn can only happen if we put ourselves in the right place at the right time. I was out there and that was the first step. In the opposite extreme, I expect to see an otter on every loch I pass in the Hebrides. I don't, but often I do, and being optimistic helps me stay alert, so that when an otter or whatever else I am watching for does appear I am prepared and don't miss it.

Judging by the many holts that I have found all over the Hebrides and west coast, by how well worn their entrances are, by the numerous well polished slipways and pathways, and by the daily fresh spraints on the territory marking posts, the otter population is quite healthy there. It is not necessary to see an otter to know if there are any in an area. They leave so much evidence behind them. Although, it is always more interesting to watch them when given the chance. Such as on one occasion while I was driving along one of the islands' meandering roads, passing along the top edge of the spring tide mark of orange kelp and black wrack, I noticed an otter swimming in the bay. I watched that otter fishing, eating its catch while treading water out in the middle of the bay and then pulling itself up out of the water onto an orange sea-weed-covered rock where it stretched out and went to sleep. All within a hundred metres of the car.

Another day, while busy collecting prey remains from an eagle's plucking post up on a coastal cliff, I looked down and noticed a dog otter come out of the sea loch directly below. It investigated the various odours lingering around a set of entrance holes to a large holt, rolled around above one of the openings then left a fresh spraint of its own on a well used marking post – a lush green hummock - before slipping off back into the sea. Such glimpses can be so fleeting and unexpected, yet in the isles they can be such an everyday sight and accepted as usual, without further thought.

I finished collecting all the prey remains and pellets that I could find at that eagle's perch. The usual species' remains, those of hare and grouse which are found at mainland sites, were few. There was only one pellet that contained feathers from a grouse as they are relatively scarce in the outer isles. Most of the pellets and feathers were of fulmar. These can be identified easily from the considerable distance of several metres by their unmistakable fishy odour. And there was a large ball of tightly bound brown hair. The hair was silky and short, densely packed into felt. It was a pellet of otter fur. How the eagle had caught the otter I do not know. It might have been dead on the shore and the eagle had simply eaten it as carrion, or perhaps it had been a small or young otter and more easily caught than a large and potentially dangerous adult. It might even have been asleep on a rock when it was snatched.

Otters have few enemies. Man is the main threat to them. He can destroy their habitat and they used to be hunted. There was once a thriving market for otter pelts in the fur trade, and any chance of financial supplement to a gamekeeper or crofter's income was taken in those days. Terriers were used by some to try to flush out otters from their holts when they could be shot as they bolted. But an otter is more than a match for most terriers as gamekeepers found after putting dogs into dens for foxes only to realise too late that the dens had been occupied by an otter. Those incidents more usually ended with the death of or severe injury to the dog. A large otter is very muscular, with a wide gape full of stiletto-sharp teeth, and otters are quick, very very quick.

Gin traps were commonly used and there is still evidence of that lying around today. The traps were set with jaws laid open at the entrance of a holt. It would snap shut on the first paw that stood on it and as it would be firmly tied by a chain to a peg nearby, the otter could be collected by the trapper whenever he next checked his work. I know of two traditional holts where there are old pegs set firm and overgrown by their doorway. And I have found several long-neglected and rotten traps lying by riverbanks. The rusted jaws were all fixed by chains to long rotten pegs. Thankfully their blunt teeth are no longer a threat.

There are traditional otter holts all over the Highlands which have been used for generations, probably hundreds if not thousands of years – for some are in natural crevices which would have been there since our shorelines were formed. Many of these old otter haunts would have been well known to the trappers. They were in all likelihood familiar with every loch, river and backwater alike. And in their searches for otters, every inch of the land and all its secrets would have been known to them. Fortunately I have seen no people on my short-cuts, and thankfully no trappers. The otters are safe.

Highland mist, a popular image of the Highlands. It can be deceptive and cause people to lose their way, panic or become totally confused. Anyone venturing out into the hills should be able to navigate with a map and compass, and global positioning system devices are a great aid. I carry all three, mostly for logging positions of my various findings, such as nests, roosts or kills. To find my way around I usually read the land, take note of features as I pass them and if I see mist approaching, I grab a quick mental picture of my surroundings before I'm immersed. Then I carry on with my work, happy and content in my own little world in the mist.

14

EAGLES IN THE MIST

Many eagle eyries are well known and have been used for generations. Some are so large they can hardly be missed by anyone looking for them. Some however, are small and set in the few quiet and secluded corners of the Highlands that are left.

Still amber light promised a fine evening. A ring ouzel was whistling from a rock on the hillside. And a pair of black-throated divers were swimming on the loch with their two grey downy chicks.

I was having a rest on the shore after a rewarding day checking the progress of an eaglet. It was in a large conspicuous eyrie that I had found earlier in the year when the adults were incubating their eggs. Now they had a healthy chick approximately six weeks old and from my knowledge of the successful history of that pair of eagles I was confident that the chick would survive to fledge. My thoughts had then turned to the neighbouring pair of eagles and how they were faring in rearing young. When I checked that home range in early spring things had looked bad. One of the two known nests had been built up a little with a platform of fresh heather and grasses, although there was no evidence of any eggs having been laid. Neither were there any signs of the birds in the immediate area. The birds might have failed to breed that year or had a nest hidden farther into the hills where I had never looked before. As they had been a productive pair up to then, I suspected the latter. The question was where to look? Daylight was well past its peak and I would have to hurry to cover the ground before dark.

Thinking was not going to get me any further on the matter so I pulled myself up and set off up an old deer stalking path and did some doing. Facing out to the thin line of the Outer Hebrides on the horizon, the glen felt clear and warm in the slowly slipping sun. Pushed for time, I left the track and took a short cut crossing over a straggling ridge which placed me well beyond the old unused eyries and onto the edge of the unknown ground. Although the walk-in distance was shorter, it proved to be heavier going, and that was my second time out on the hill that day. The steep climb had left me tired. I slumped onto a prow of rock and leaned on my rucksack. There were hills and cliffs spread all around me. An eyrie could have been hidden on any one of them. I spied what I could through the telescope, but my view was too limited by the knobbly landscape. Spying through the telescope has often saved me miles of walking at other sites; not this time however, I could not find a clear view of all the crags. I was going to have to walk all around all the hills until I found something or proved that there was nothing there.

A long chain of ridges linked by rocky gullies strung out away from me. Above these cliffs there were numerous others of various sizes scattered across the broken hillside. And down below, dark shadows hid smaller hillocks that had been cleaved to create perfectly suitable cliffs for an eyrie. I set off again, following the crags, checking them one by one. Lochs and lochans filled the floor of the glen, and widely strewn rocks – big rocks, the size of houses - added to my meandering course over the boggy moor. Deer trails helped my progress in some places, rank heather tripped my ankles in other places. The peat was wet and heavy, and pools of yellow sphagnum moss waited to pull off my boots. There were very few signs of animal life. Four hinds with two calves barked and bleated as they trotted upwind away from my scent. A solitary wren clicked alarm as I passed a shallow gorge where she probably had a brood of chicks. Excepting these, only the repetitive squeaking of meadow pipits followed my steps. There were few trees. The eldest of them bent crooked on the inaccessible crags or lay low on a tiny islet in a lochan. Tiny insectivorous butterwort and sundew were the dominant flowers in the more open parts of the nutrient-poor moor. Yellow bog asphodel and tormentil added most colour to the bleak brown landscape.

The first cliff that was potentially good for a nest was no more than potentially good, neither was the second, nor the third. In the next gully however, I found a very old ring of sticks high up on a ledge, and overgrown by a tangle of grasses, heather and foxgloves. It must have been a historical site, one which the eagles might come back to some day. There were no signs of any eagles having been near the area recently, so I logged the co-ordinates and kept on walking. I climbed up a buttress and scrambled down a gully. The rocks were thickly vegetated forcing me to use heather and woodrush for handholds as I pulled up and leaned over cliffs for a quick look here or a last search there.

Two eaglets, about tens days old, typically sit far apart within the eyrie. The stronger bird occupies the centre of the nest with its head upright while the weaker one lies with its head down at the edge of the nest keeping out of range of any strikes from its sibling. Although there is plenty food in the nest there is still a strict hierarchy.

While grovelling like this I was also hurrying and I had no idea how long I had been doing this when an eagle appeared high over a ridge another two kilometres farther on. My head had been down so I never saw exactly where the bird had come from. It was flying slowly and rising on widespread wings, so I suspected that it had just taken off, and I was hopeful that it had just left a nest or at least a favourite roost site. At last I was in the valley where, if there was an active nest, it would most likely be.

The bird wheeled round and round in overlapping circles until it had climbed high up over the hillside to my left. In less than a minute that eagle had cast another full kilometre between us before slipping around the backside of the topmost ridge. When I first saw the bird I fooled myself to believe that I had found the nest site. Eager to succeed I had ignored the simplicity of the meaning of sighting the bird, conning myself. The only certainty was that there was one bird in the territory at least, no more than was known earlier that year when I first visited the area and found signs of eagles on the ground. There were still several cliffs to check and time was running out as the evening wore on. I hitched up my rucksack and set off once again, following a methodical pattern up and down, around and across. Even if there was nothing to find, it had to be proved, and proving a negative can be the hardest task of all. Although eagles can and do build very large nests, they do not always do so. Some nests can be quite small, a simple ring of heather, and well concealed on nothing steeper than a heather-clad slope. All ledges had to searched.

And then, to make things even more difficult, a thick mist crept up the glen. Silently, fingers of grey chilling brew ran through the gullies and out over the hillside. Down below on the moor, the lochans lost their shape. Only a mirrored glare revealed their position. The fog had caught me unready as it rose up from the cold waters of the sea loch. Even the bright setting sun was so thickly and quickly obscured that my sense of direction was knocked out. I sat down on the spot, dejected and weary, but stubbornly determined to see the job through.

A jumper pulled from my pack stifled a shiver. I knew where I was and where I wanted to be, only the details on the map and simple compass bearings were not quite good enough for following a route up and down and round about all those hillocks and hollows. In compromise, I used the map to identify the hill that the bird had most probably risen from. That told me the distance and the direction separating us. How much light was left? Not much, although there was maybe enough in such a long mid-summer evening for the distance involved. The thought of being so close, to leave, and the effort involved in coming back another day was too exhausting to accept. One chance - take it.

There were two kilometres between me and the cliff as the crow flies - above the mist. But, I was deep set in the mirk. As I approached grey cliffs hidden in the grey mist they shrank in their materialisation to insignificant boulders or bluffs.

A rocky promontory above a corrie in the west Highlands where golden eagles often sit.

Eagles often sit on the skyline, watching all below. However, unlike as so often portrayed in dramatic illustrations such as atop sharp pointed pinnacles, their perches are usually mossy or grassy topped rocks. Bare hard rock would blunt their talons.

Midsummer sunset from the summit of A'Mhaighdean in Fisherfield.

There are more than twenty hours of daylight for eagles to hunt prey for their half-grown chicks
- more than enough time.
The numbers of hares, grouse, ptarmigan, plovers and other assorted prey are at a breeding
season high.
- a time of plenty for eagles.

After clambering around or over several of such obstacles and circumnavigating a lochan, I doubted if I was still on course. Even by following the compass it was difficult to counterbalance anxious uncertainty with confidence. Judging the distance travelled was the main difficulty. How far had I gone? Had I overshot my target? When should I turn back? (This was all long before personal geographic positioning systems were available).

My eyes were peeled and my ears were pricked. Every so often I stopped to glance up or cock an ear. All that I could see was heather and mist. And the sound of the mist rolling through the heather tormented my ears with its spectral whispers. Then there was a cheep. Then another. It was a thin ethereal call, and I was sure I knew what it was. The call was repeated again and again with distinct excitement, and I was reassured. It was the call of an eaglet begging for food from a parent. Now, where was it? How far away was it? Such a call could easily have carried for hundreds of metres in that still atmosphere. I had achieved what I had come for, yet I could not put my finger on it. It was all so tantalising. And the calls were becoming more excited. An adult was definitely close to the nest. My stomach drew in and my neck stretched out.

Stealthily, I crept on into the mist. My only beacon was the chirp and I was wary of every shadow I peered into, careful not to disturb the birds at dinner. Then there it was. I sank to the ground, then peeped over a small hillock to watch the birds on an eyrie about one hundred metres away. The ground was wet, so I shifted in behind a large boulder. Leaning against the rock, I pulled out the telescope and spied at the nest. Luckily, the birds had not noticed me at all, the adult had her back to me, and the chicks were too busy in their clamour for food. The hen was standing firm on the edge of the eyrie, a scant construction set on a broad fern-clad ledge and tucked in behind a knarled old rowan. She was pulling and tearing at a piece of prey held tight beneath her great yellow talons. Her deep curved bill tore at the flesh. Yet the same savage bill delicately offered tiny stripped morsels to the outstretched gapes of two chicks sitting in the centre of the nest.

Still mostly white in their warm downy plumage, the chicks were about four weeks old. They held themselves upright while squatting on their heels. One was noticeably larger than the other and it sat slightly forward of its sibling. That one's crop was bulging and it grabbed most of the offerings from the adult. The smaller chick's crop was near empty and it was repeatedly and viciously pecked on the head by its larger sibling. As the adult meticulously butchered the prey I could identify it as a fox cub. It was still sooty in colour with a flash of red on its legs. The adult bird towered over her chicks. In contrast to their apparent frailty, she was a solid form. Her broad strong head was spattered with blood. Her wings were held back tightly as she pulled at the cub's body, and the long feathers on her legs hid the strength of her grasp.

Unaware of their secretive audience, they enthralled me for three-quarters of an hour. All that time wisps of mist would repeatedly veil and reveal my view. Then a chick looked skyward to my left and called. It was watching something and it called again, this time more frantically. I glanced up over the rocks behind me. And there, skimming past the lip of the cliffs, was the male bird. A second fox cub was firmly fixed in his talons, its weight stretching the bird's legs and keeping his flight low. The male, who in most cases does the majority of the hunting in the early stages of the chicks' growth, had obviously located a fox den somewhere in the hills. Having successfully caught one cub previously, he had probably returned to the same area to harry the rest of the litter. It was likely him that I had seen earlier, leaving the nest cliff. He would have delivered the first cub for the hen to distribute between the chicks while he went off for more food. He had done well.

One unfortunate vixen had now lost two of her cubs in one day. I have seen evidence of other eagles having done likewise before. And the cubs have all been the same age as the ones I saw that day. That is, about a month old, when they begin to explore the world immediately around their den. As golden eagles are opportunist hunters, such easily caught prey is often exploited. Other examples of seasonal prey are raven, crow and heron nestlings and red-breasted merganser ducklings. As some gamekeepers consider all these as vermin, the eagles must be helping them. Yet some gamekeepers continue to kill eagles. Surely the natural balance is best.

The male eagle cruised over the nest cliff to a chorus of chirps and yelps. He turned and landed on top of the nest ridge, where he appeared to wait for the female to come and collect the prey. Well aware of his movements and of what he was carrying; she stopped feeding the chicks and left them to pick at what was left of the carcass between them. She turned off the eyrie and floated on the soft misty updraft to land beside her mate, then dipped behind a hillock out of my sight. Light was failing fast so I waited for another ten minutes. She never reappeared with the food, she probably ate it herself, for the chicks had been well fed, or rather, one of the chicks had been well fed.

Another long day was almost over. I used the last of the twilight to guide me back through the mist, and my pace was speedy and direct in fear of benightment. When I at last stepped onto the old stalkers' path, I sighed in relief. It meant reassurance of finding my way back off the hill, for it was well dark by then. From there, I left my legs to walk on back on their own, while my mind went back for another quick squint at those boisterous chicks in their secretive eyrie.

When I checked the nest site about a month later to see if the chicks were ready to fledge my fears for the smaller chick's safety were confirmed. One chick was large and almost ready to fly. Its crop was full again and there was prey on the nest beside it. The smaller chick's body was lying in the gap between the eyrie and the cliff. The skin around the head was all torn and the flesh bruised. It had died perhaps a week after I had watched them. Siblicide by eagle chicks is quite common, but it seems so needless when there appears to be enough food for two in the nest. The amount of food brought to the nest by adult eagles for their young is critical in determining how many chicks can survive to fledge. However, there is more to eagle chick survival than just food abundance. Even when there appears to have been enough food for two chicks, one can die. One chick might beg more persistently or be more aggressive; is there enough food or is that simply our impression: the reason has yet to be unravelled.

The chick in this photograph was particularly aggressive to me as I approached the eyrie. Its sibling, which I saw being pecked in the head a few weeks previously, lies dead at the back of the nest.

The spectacular open hillsides associated with the Highlands today are a poor remnant of their former glory. In twelve thousand years the vegetation cover has changed from post glacial tundra, through scrub cover and expansive woodland; then the woodland receded as the climate became more oceanic, and open moorland has increased since man in the bronze age began to extensively clear woodland and scrub. At its climax the plant cover of the Highlands would have been a wonderfully rich mosaic: with tall thick woodland in the richer soils on valley floors, especially in the south and east; stunted scattered trees in the bog; juniper, birch and willow scrub mixed through the heaths; and up on the windswept ridges there would have been short heaths and grasslands. The oceanic climate has favoured the growth of peat, especially in the north-west, which in turn has restricted tree growth. The familiar sheep grazing in the Highlands is only about two hundred years old. The older, more traditional hill stock was cattle with some goats and sheep. And all these domestic animals would have inhibited tree and scrub regeneration. Now wild red deer are also degrading the landscape, due to man's inadequate control of the deer population since his elimination of their main predator, the wolf, and the encouragement of high numbers of deer for sport. Golden eagles hunt over open hillsides, so all these causes of increase in open ground would have augmented the eagles' habitat, so long as there is adequate prey, and Scotland holds some of the highest densities of golden eagles in the world.

15

THERMAL OF BIRDS

Two pairs of ravens, two pairs of buzzards and a pair of golden eagles all in one spiral.

The northern Highland landscape is not so much one of high mountains, as one of rolling moorland with a scattering of high pointed peaks. There are numerous hags in the peaty blanket bogs, and lochans, pools and meandering streams. The main birds of that type of country are golden plover, dunlin, greenshank, skylark, and meadow pipit. Hundreds and thousands of pipits. There are also a small numbers of red grouse mountain hares. And there are hundreds of red deer.

Early one morning, I was up on top of a hill looking over a wide panorama out to the west, when I heard a pair of ravens calling excitedly. From my experience of raven behaviour, it was likely that they were up to some mischief, possibly harassing an eagle that had made a kill. The ravens were deep in a gully out of sight. And as the fuss died down, I continued to spy around for eagles, or any other birds, which might be in view from my vantage point.

The day was warming up and the sun was out, not particularly good for spying as there was a thick wobbly heat-haze building up. There was a herd of red deer down on the moor below, a huge wide moor, stretching for miles into the distance. Most of the deer were hinds and yearlings. The new calves were due to be born any day; although I had a good look I couldn't see any. And the few young stags that were in amongst the herd were only just discernible from the hinds in the poor light, as their antlers were less than half the size of an adult's mighty spread. The whole scene was reminiscent of an African plain, with the grazing herd slowly walking along. And the deer were obviously feeling the heat, or being pestered by flies, as they began to collect in a tight group in the middle of a shallow lochan. Their images flickering through the haze as if in an old shaky film.

A wee while later, a pair of ravens lifted out from a gap in the ridge about a kilometre away. They were quickly followed by a second pair of ravens, and then two buzzards, again a pair, and then another two buzzards, which proved to be a pair also. And after all that, two golden eagles came up - their wingspan dwarfing those of the others. All ten birds flew in a tight spiral and up on a thermal of warm air rising from the sunlit southern slope of the ridge. The ravens were calling again, and flew quickly, jinking about, mobbing the others. The buzzards soon caught them up as their wings are better shaped for soaring. Before long, the two eagles were in amongst the buzzards and all six raptors were in one big wheel about one hundred metres above me.

The ravens never gained the same height as the other birds. One pair of them sloped off to the north over the hill and down a burnside, pruk-prukking to one another as they went in their tumbling raven flight. There were plenty of small cliffs on the northern slopes for them to nest on, although they probably had free-flying young to tend to at that time of the year. The other ravens flew back into the gully, probably to have another scavenge for scraps – for there was definitely a kill lying just out of my sight. They came out again several minutes later and flew off towards the west – to deliver some food to their young?

All four buzzards left the thermal when about two hundred metres up. They separated into their pairs, and one flew down each of the burns running south on either side of the deer herd. There was a long stretch of continuous woodland filling the steep slope above the river. Most of the trees were birch, with a few jagged geometric stands of pine plantations. Either would be good enough for the buzzards to nest in, and I found a buzzard nest, in a birch, on my way off the hill in the afternoon.

Meanwhile, the eagles continued to soar on up to over three hundred metres above the hilltop and only then did they take one long glide to the east. I knew where their nest was, and they had no young. They had failed in their nesting attempt for that year. Neither of them was carrying food in its talons. They would simply have eaten what they could from the kill, and were now casually flying back to the centre of their home range. As they passed over a ridge they dipped into their nesting corrie, dead on course for the main roost by a group of eyries, where I thought they would sit quietly for the rest of the day.

Later, more than an hour later, one of the pairs of the buzzards came back – and one bird flew away with something small and furry in its bill. Then a little later, the ravens came back to scavenge again, but didn't seem to find anything as they soon flew away and continued to forage elsewhere on the moor. Finally, an eagle flew through the gap – checking for more prey? Yes, I could work out what had happened now.

And I was right, as I proved when I checked the scene of the kill later. At first, I had thought the prey might have been a newly born red deer calf, and then I found a fox den that had fresh paw prints around it. There were no eagle prey remains around the entrance, just a single eagle feather. An eagle can eat a fox cub without too much butchery. It could have torn it into quarters and snapped off the sharp claws which could damage its crop. The young bones would be very easily snipped by an eagle's powerful bill, and the flesh just as easily torn between bill and talons.

It had been a fox paw that the buzzard had flown off with and that was probably all that had been left of the feeding frenzy. I couldn't find any scraps at all. The scene had been picked clean. The single eagle flying low through the gap afterwards, had probably been one of the previous birds taking a quick fly-past, checking for any other cubs that might have come out of the den to play. A few hours had passed since the kill and commotion. No luck that time though. The cub that had been killed had probably been lying sunning itself at the mouth of the den which was set in a sandy hillock with a dry heather bank. That is a typical situation for a fox den in the Highlands, and this one was obviously a traditional site that had been occupied for possibly thousands of years. There were five entrance holes, some in use, and others overgrown. Foxes had probably dug their dens there since the end of the last ice age. Perhaps wolves used to use it.

The sandy hillock was a deposit left in the mouth of a glacial meltwater channel, which had carved the gap through the hills. Such sites are common in the Highlands where during the last great ice age, large glaciers covered most of the land. The den was in the east-facing slope, sheltered from the prevailing westerly winds. The sand was a dry island in a sea of wet peat. It was a warm, dry comfortable spot, but it was exposed to eagles. When the hillsides had more wood and scrub cover it would have been a much safer place for the foxes.

I had just witnessed as good an exhibition of animal behaviour as any which people are enthralled by in images of the African plains and its wildlife. Right there in Scotland. And then I thought of the wolves. It would have been fantastic to have sat there on that exact spot and watched a pack of wolves make a kill from the deer herd, with the eagles, buzzards and ravens scavenging the wolves' leftovers.

When ringing eagle chicks, as with other raptors, it is not so much the large hooked bill that needs to be minded, but the massive, powerful and needle sharp talons. Their toes are as large as our fingers and have a true vice-like grip. It is the constriction of that grip which kills most of their prey; the accuracy of a talon piercing an important organ would be too much of a chance in a hunt. The main function of the claws is to grasp prey tightly while the eagle hangs on long enough to overpower it.

16

RINGING CHICKS

After watching eagles from a distance all year, there is a great feeling of attachment when handling the young birds while ringing them, and satisfaction when they successfully fledge.

An important part of the study of golden eagles involves the ringing of birds, especially the chicks, to help in our knowledge of where the birds disperse to from their natal site. And in time, we should gain an insight as to just how long these commonly regarded as long-lived birds do actually live for.

The chicks are best ringed when they are about five or six weeks old, as by then any which are going to die for whatever reason, such as siblicide, will have done so. It is rare for older chicks to die. The chicks are usually that age in June and early July, a time when most of the difficult fieldwork has been done and all that is left to do is follow up the active nests and complete the season's records. Fortunately, by then the cliffs are usually reasonably dry and safe to climb on, and there are the long hazy days of high summer to work in.

Some eyries are easy to access. It is possible to actually walk into them and sit by the nest. However, these are becoming scarcer as they are not only popular with ornithologists working on eagle studies, but also with the public. I know of several sites that have become tourist eyries. I have heard of people being sent to an eyrie with explicit instructions from a hotelier. And at another nest site, a large patch of flattened grass indicated where people had a picnic

Even as youngsters, golden eagles have piercing, concentrating eyes. They watch everything around them and are quick to react. Here a chick prepares to attack by pecking with its well developed and very sharp bill, although they mostly use their talons to attack. Over the years I have found that it is usually female chicks which are more aggressive, the males tend to lie quietly in the nest even if it is a lone nestling.

while at the nest; leaving film packets, drinks cartons, sweet wrappers, and all sorts of litter lying around the base of the cliff. No wonder such nest sites are becoming abandoned by the birds. I always dash in and out of a nest area to impact as little disturbance on the birds as possible, and I am licensed to do so as is required by law. Disturbance of rare birds is a serious offence, there simply isn't enough space or enough birds for anyone to cause unnecessary disturbance.

Licences are issued by Scottish Natural heritage, and special photography licences are also required before anyone can legally photograph adult eagles at the nest. This involves using a hide and I have never had time to do this properly so have never done so. Most photographers do abide by the rules, but as always there are the selfish and inept. They are usually male, as are most trophy collectors, for after all that is what most of the photographs are. I have known a pair of eagles to abandon their nest and eggs because an unlicensed photographer erected a hide too close to the eyrie. I suspect others of having done the same, and I've found other hides at nests which shouldn't have been there.

Egg collectors (they are invariably men) are another problem; they cause senseless harm to our eagles. There might have once been an arguable case for egg-collecting when much science was based on early findings, there was less understanding of avian ecology, or because the birds were not endangered. But there is no justifiable reason for it today.

I have found such evidence as footprints, torn vegetation, wooden stakes and ropes on my way into empty nests. The covetous behaviour of ill-informed tourists, photographers and egg collectors is intruding on the eagles, both the golden and the sea eagle. The selfish are spoiling it all for themselves just as much as for the eagles and for those who care about the eagles and their future in our hills.

And it is not only deliberate disturbance that is affecting the eagles. Over the past thirty years numerous eyries have become abandoned, probably for a different reason, they are too close to roads or footpaths. Some paths in the hills that were seldom trod on thirty years ago, can now be used by hundreds of people in a weekend. Events such as sponsored walks or large group outings are particularly bad for this. Any mass passage through the wrong place in spring could result in eagles being kept off their eggs and the embryos dying. In some instances, routes that are too close to eyries are promoted in guides. This unintentional disturbance is a consequence of the recent upsurge of people's interest in the hills and an ignorance of what is there.

Fortunately there are still some quiet corners for eagles to nest in, and some rather grand cliffs where they should be safe from most disturbances. As I usually work alone, I don't take risks climbing into such nest sites. However, it

A fly crawls between an eaglet's ear and eye – they continually pester eaglets in the nest on warm summer days when decaying prey remains attract them. There is always some residue of blood or food stuck to their bill and claws and the adults are irritated too. I often wonder if the adults deliberately soar high on such days to escape their annoyance.

is surprising how many can be accessed with a little cunning use of ropes, and careful study of the cliff from far off to pick out a route through the ledges.

Climbs into eyries should be uneventful, focused and efficient. There is work to do. And I find it easier to do it alone if possible, for the least disturbance the better. If a rope is required, a helper is often required to hold the rope. Other eagle workers help me on such occasions, and I help them when required. However, help is not always at hand. One such occasion began with a walk up and into the higher end of a nesting glen. I passed through a scattering of large boulders - glacial erratics dropped by the ice more than ten thousand years ago. Then as my view cleared, I immediately saw an adult eagle turning round the ridge where the eyrie was, and I could even see the chick on the nest. The eyrie was by then well splashed with white droppings and dotted with cast white chick down. The adult bird slipped away to watch my proceedings from a safe distance on a perch on the opposite ridge. A kilometre or so away.

I spied up at the nest to check all was well, and sure enough, the chick was sitting upright watching everything around. No doubt, including me. The nest site was in a corner under a huge overhang of rock, about three metres deep. And the nest was an equally large haystack affair. As I approached the foot of the cliff, the black roof extended out into the blue sky behind and my own size came into perspective. I had free-climbed into the eyrie from below several times before. There are a few high steps over a small overhang at the base of the cliff, followed by a sequence of less confident holds on grassy ledges leading up to the bottom of the nest. Then I usually squeezed up a chimney formed between the nest and the rock behind. The climbing would be technically graded as very difficult, which despite the wording, is easy by today's rock-climbing standards. Unfortunately, the crag was wet that day. The rock was slippery, especially where it was whitewashed with the birds' droppings, and the turf could not be trusted to stick to the ledges when so waterlogged. So, I had to go in from above on a rope.

I clambered up and around the back of the crag to find a position above the nest where I could rig up a belay. The rock was lewisian gneiss, which is a good solid type with plenty of cracks to fix gear into and boulders to slip slings around. I attached the rope to two good solid anchor points, lent my weight on the rope to test the fittings, flung the loose end down the face, and then slowly walked over the edge. Abseiling in real life is completely different from that seen by most people in movies and charity stunts. This was no safe and clean stage set. There was no room for movie heroics. The trick is always to do it safely and properly, with both me and the birds in mind. Loose bits of rock lay on ledges and in cracks, so I had to take care not to knock any off with my feet and send them crashing below. And even more care had to be taken not to knock any off with the rope above as even a small chip of rock can seriously injure when speeding down onto one's head. I always wear a helmet when abseiling, but there are plenty of other parts of the body to hurt.

I eased down slowly over the vegetated gravel on the lip, lent back as I walked down a straight wall section and then swung in under the overhang and sat there in my harness, feet dangling in the air and spinning slowly with the rope, eye to eye with the eaglet sitting on top of its nest. It hissed and flapped its stumpy wings. But it was, and always is, the amazing eye contact which it held on me that was the most frightening. We humans shy away from such strong eye contact. We cannot hold it for so long with such intensity. I refocused on the rope situation and swung to catch hold of an edge of rock under the roof. Then I gradually eased myself into contact with the rock on the side wall of the corner and nudged onto the front edge of the eyrie, with my legs spread wide to either side. One foot found a tiny bulge, enough to ease my weight between rock and rope. I tied off the rope so that I couldn't slide down any farther and freed my hands for business.

There was the usual dusty air of down, flies and rotting flesh. Speed was of the essence for more reasons than one. I gently scooped up the chick to keep it close to me, and safe from falling off the far edge of the nest. There was a bit of a wobble to the nest, although all was safe and secure. It is while sitting on eyries like that, that I always marvel at their size and structure. I know some are big, really big, and I have been into some eyries that are wide and strong enough to support two people and two well-grown eaglets with room to spare. They really are amazing structures. The layers of different year's nests are clearly definable, and there are sandwich fillings of bones in between some. That nest had been wedged into its rocky cleft for years. Eventually, the old sticks rot and fall out, and whole eyries do come down, especially in winter when they are waterlogged or covered by deep and heavy snow. Then the birds simply get on with rebuilding them as they have done for centuries.

I set into the routine of firstly fastening a ring around the bird's leg. Then I grovelled quickly through the nest debris to count odd bits of fur and feather from the prey, and collected any pellets and cast eagle feathers. The chick, like most four-week olds, was easily held and, by the small size of its talons, it was thankfully a male. I say thankfully because they are usually the more docile sex, and from my experience, have a less painful grip than female eaglets which tend to slash and grab more. The first things to catch hold of when handling all raptors is their feet as these are their main weapons. Young talons are needle sharp and can pinch deep into flesh. It is important to always be up to date with tetanus injections when working with raptors. Who knows what is injected into us when we get gripped by those needles? And to top all that, female eaglets will also try to bite, and they have rather frighteningly large and strong bills – especially so when seen very close up.

The third, nictitating eyelid flicks over an eaglet's eye - the photograph was taken at 1/800th of a second. This eyelid is translucent and slips horizontally across the eye, while the upper and lower eyelids slip up and down. Its purpose is to remove dust or other small debris from the eyeball surface and they are used rapidly when raptors dive through the air.

After I had scavenged the eyrie I lay the bird at the back in a little niche to let it feel secure and less likely to jump back out. Then I undid my safety knot and slid back down to earth. All that was required was a quick scramble back up to the top of the cliff, pull up the rope, collect all the gear, and dash off across the hillside. And as if from nowhere, the adult birds both reappeared in the sky and watched me as I retreated out of the glen.

That bird has never been found and the ring number read. It might still be alive, or it might have died in its first year, soon after fledging, as that is the time of greatest mortality in eagles. However, it certainly did fledge as I checked the eyrie at the end of the summer and the chick was sitting on a roost cliff on the other side of the glen. Exactly how young eagles disperse and gain occupancy of their own home ranges has still to be unravelled. And ringing of the birds is only one method to help answer the question.

There are few records from birds older than two years old, and as many of the rings that have been recovered were from birds killed deliberately by man on ground managed for sheep or grouse, few birds would be expected to attain any great age in such an environment. Eagles are likely killed off in such places as soon as they arrive there, on replacement of any eagles which previously occupied these areas of otherwise suitable habitat. Any old ringed birds are probably living out their full and natural lifespans in ground more favourably managed for eagles.

Most ringing of eagles in Scotland has been done since the late nineteen-seventies. Other studies have shown that golden and similar species of eagles can be expected to live for between about ten and thirty years, so there is time yet to find ringed birds that have died at an older age. However, few dead eagles are found, and I have never found any other than those that have been killed by man.

As I packed up the belay after ringing that chick I found several more pellets and eagle feathers. The adult birds obviously had a habit of sitting on the cliff top ledges for hours watching over the nest and the glen below. All the while they would have been preening their plumage which is so important to all birds if they are to fly well.

Most eagle work is done remotely through binoculars, telescopes, radio tracking and feather collection. The best information comes from undisturbed subjects and we would not mark the birds if they were unlikely to behave naturally. Any bird that is ringed or tagged, and they are few per annum, is only ever handled once in its life and briefly. Yet from such minimal contact much information on eagles has been gathered and with further development of modern techniques I feel that we will soon have a more true assessment of the current status of golden eagles in Scotland.

Results from the recovery of ringed golden eagle chicks (they have mostly been found dead) indicate that they disperse in any direction from their natal site, and the birds which were younger when they died seemed to have travelled farther than older birds. This suggests that most birds initially set off on wider explorations before returning to within the general area around their parents' home range.

One new method is to attach satellite-received radio transmitters to fledglings. This can give more information more quickly and allows almost real-time tracking of Scotland's eagles as they move around the country. These are small devices that are fitted onto the eagles with a backpack style harness – specially designed to fall off within a few years. In that time the locations of the birds can be logged into computer files and the information downloaded at ease in the office. The initial results support those from the ringed birds' movements, for tagged birds have indeed travelled in all directions and some have returned close to their natal area. Further data from the tagged birds also gives previously unknown information on where the birds have been in between. In a few more years we should have more understanding on how eagles disperse and what habitats they seek or avoid.

A rigorous scientific method for identifying individual birds is the use of DNA analysis, which can be done with samples of feathers or mouth swabs from birds handled. A database of eagle DNA patterns is now being established, the samples being collected by a network of amateurs and professionals who study eagles, and the whole is co-ordinated by the Scottish Raptor Study Groups and Natural Research Ltd. Soon we will gain such information as paternity, natal dispersal, recruitment and lifespan.

Any sizable cast feather can be collected and sent off for DNA analysis; primaries, secondaries, tail feathers, coverts, even small body feathers can all be tested.

Breakfast for a prince of birds.

This eyrie was visited before eight-o-clock in the morning and there were already freshly caught items of prey presented to the eaglet - on top of what was already in the nest from previous days and still perfectly edible. The total list of identifiable prey was: one ptarmigan, one raven, two fulmars, two red-breasted merganser ducklings, parts of two mountain hares, and part of one red deer calf. At other times when I visited the same nest there was no prey, so this scene was not typical.

Adult golden eagles bring two or three food items to the nest per day, weighing about a kilogram in total: of which a well grown chick will eat about half. The adults take away and eat any excess.

17

PREY

Golden eagles, are at the top of the food chain in the Highlands, and their broad diet includes numerous epicurean delights.

After the chicks have fledged in July it is time to thoroughly investigate what prey the eagles have been eating by climbing into and rummaging through their eyries. In the eastern Highlands the main items of food are hares and grouse with little variety. In the west, where these animals are scarce, they take a wider selection. And it is interesting what eagles will take as prey in these areas and how their diet varies between regions according to what is available in the local habitats around the nest sites. Man's use of the land has a great impact on this, but eagles are versatile.

The landscape in the west Highlands is one of long rugged coastlines running around lumpy peninsulas, the spines of which are formed from long twisting ridges, jagged in profile where cliffs and gullies contort the contours. There are some fine woods of oak, ash, pine and birch. And steep slopes plunge straight to the very banks of the lochs, rivers and burns, leaving scarce enough level ground by their shores for a shieling. Most settlements are small and scattered thinly around the coast and headwaters of the many sea-lochs. Similarly, the roads hug the coastline, only meandering over steep passes now and then. And within the rings of roads there are large tracts of open hill that are grazed by sheep and red deer.

High rainfall in the west, which encourages mosses to grow and peat develop, inhibits heather. It is more the grasses that thrive under the precipitation. So there are fewer grouse and hares than in the drier, more eastern hills where hares and grouse thrive in the abundant heather. In spring and summer, the lush green sweeps of the west's hillsides are bright to eyes that are used to the drier heaths of the east. The rainfall is also good for exotic conifers, and in the south-west particularly, thousands of hectares have been fenced off and deep-ploughed for forestry. The land has been planted with alien tree species and massive monetary monocultures have choked the life from the land. Pine plantations have smothered so much open hunting ground in some areas that eagles can no longer live there. Eagles can hunt through open woodland where there are roe, hares and black game, but not where trees are too tight together to fly between and the ground below holds very few prey.

There are some glens that are filled to their skylines with conifers. And in amongst these there are small pockets of open space where the hill is too rough, or too steep to plant, and on the wind-blown summits - barely enough to let the land breath. Within one of those pockets there is a cliff which used to be a traditional nest site of golden eagles. The eagles are now gone. The nest ledge is now used by a pair of ravens. Another similar site is now occupied by a pair of buzzards and another by a pair of kestrels. And there are more similar examples. There is no longer enough open heath and associated prey species available for the eagles to hunt. So, eagles are now extinct in such areas. The smaller buzzards, ravens and kestrels can live on the voles and invertebrates which can still be found there. That is not enough to sustain eagles.

Fortunately, in recent years there has been some re-shaping of these plantations as they have reached maturity, been felled and re-planted in a sequential pattern that should open up part of the woods at any one time. Hopefully this will be enough for small mammals and birds, and the eagles may once again hunt over the ground. And there are still vast tracts of land that have never been planted. There might not be much food there compared with the eastern moors but eagles seem to find enough to rear chicks.

Summer is well on by the time the chicks fledge. The warm sunny days are a treat and the walks up the glens are a stroll. All in big contrast to the last cold stormy days of winter when the eagles first began to display and build their nests in the grey sodden hills.

One such day, while I was with Jeff Watson, we went to check a nest in Lochaber. Jeff knew that a chick had been successfully reared at the site we were going to, so off we set on a fine August morning. It was good to have company for once, after so many days on the hill on our own. Our way led around the fringe of the hills by a sea loch where there is a narrow belt of oak wood. They are grand old trees with thick heavy bows, and the woodland is rich in lichens,

Eagles, like all birds that eat solid food, cough up the toughest and more indigestible parts of their food in pellets. With golden eagles this usually consists of balls of hair, feathers and bone. On inspection of the above pellet by partially breaking it open, the skull of a stoat was clearly visible and identifiable, complete with a large hole to the rear of the skull showing where the eagle had nipped it and dealt the death blow.

mosses, ferns and fungi which entwine their own miniature jungle on the ribs and knobs on the trunks and branches. Farther on there were fine stands of birch, some rowan and a sprinkling of hazel, aspen and holly. Honeysuckle blossom hung from a small crag. Foxgloves and hawkweeds paraded the ditch-side. Beef cattle browsed knee deep in wet meadow grasses and flowers, keeping the glades open. Electric blue damselflies shocked the still air between the trees, and a cock redstart flycatched from a stump.

It was good to see such a variety of wildlife in such a short distance. And there was more as we came out of the trees. A stonechat sat erect on another stump, above some tall bracken where his brood lay hidden, his alarm call sounding so much like a stone being repeatedly rasped on steel. And a buzzard's mewed alarm came in from above the first of the cliffs.

A waist-high sea of bracken fronds filled the shallow glen. So we climbed up onto a ridge to find easier ground, tripping over roots and stones unseen on the seabed. A roe's white tail bounced through the green waves. And as one animal ran away from us, thousands of others rushed towards us. A smothering cloud of flies crowded around me, crawling into my eyes, ears, mouth and nose. Clegs followed on, landing softly on any exposed patch of skin then sticking sharp red-hot needles into my flesh. Sheep ticks streamed up my legs to suck my blood. And midges swarmed around me whenever I stopped to shoo off the others. As we reached the high ground, where a cool breeze eased the strain and suffering of the insects attentions, I thought warmly of the cold spring days when there were no pests around.

Having escaped, still with some blood, we stopped to plot our route through the innumerable crags that were set in front of us at all different angles and heights. Some held remnants of the native woodland; others were terraces of grass divided by numerous sheep paths. We flushed a pair of hoodies off their nest cliff. The adults created a fuss, while their youngsters skulked craftily in the thick cover of wild rose bushes and ivy that hung over the face. They kept quiet to avoid attention until their parents gave the all clear after we had passed and then all five birds sneaked off round a corner.

On the next cliff, a brood of kestrels struck up a thin shrieking, although not for long. The young birds were well fledged and flipped over the cliff top, soaring back over now and again for a look at the intruders, no longer alarmed. A telltale streak of white droppings and grey down from a black cleft in another cliff told of the successful rearing of a brood of raven chicks. And they hadn't strayed far either. We found the whole family not five hundred metres away, sat on some hummocks and rocks on the top of a knoll. They were bouncing and cavorting in a mischievous dance, watching every move that we made, yet keeping a safe distance.

Several knolls and gullies and ribs later, we at last looked down on the eagles' nest cliff. An adult bird immediately passed low along the top of the crag, then disappeared not to be seen during the rest of the time we were there. This was a typical fly-past which eagles usually make when their nest is approached, and it is a good indication that the birds are breeding. Although it is easily missed. Sunlight shone through the gap in the eagle's wings where flight feathers had been recently moulted out. The first primary feather was missing from each wing and the sixth was half regrown. Golden eagles mostly cast their feathers during the summer when there is plenty food. They never completely replace all their feathers in one year, as it requires a lot of time and energy to do so. The summer is short, the feathers are slow to grow, and eagles need to remain in good flight order to catch their prey.

At the instant the eagle showed its form in the sky a loud whistling cheep came pleading from below. A fledged chick was sitting on a knoll begging for food from its parent. Its calls were ignored, then it looked our way with an enquiring eye and flapped clumsily and weakly away to a perch at a wider distance from us. Resettled, it bobbed and twitched its head as it watched us then craned its head back and high to follow its parent's desertion.

Although the nest was now empty, we still spent as short a time as possible at the scene in order to lessen any disturbance of the birds' secluded refuge. We quickly rigged a belay and I scrambled down through the tree-clad cliff. By that time in the season, the eyrie had been trampled down by the birds to an untidy tumble of sticks. Dusty feather scales and cast down pillowed the dry faded grass lining. A withered spray of rowan leaves lay curled up on the rim – the last of the greenery to have been added that year. The nest was precariously balanced on an incommodious ledge, and the whole construction wobbled as I eased part of my weight onto it while still attached to the rope.

There was a fair selection of fur, feathers, bones and assorted anatomical parts wedged in the gap between the nest and the cliff. Every item was collected into a bag for less rushed identification later. Ensnared in the matrix of the nest sticks was a collection of pellets, the regurgitated castings of indigestible fur, feather and bones, which are swallowed with the meat when the birds feed. All of these were collected in another bag for precise analysis in a laboratory later once dried. When I was satisfied that every scrap of evidence had been gleaned from the scene, I made my way down to the bottom of the cliff directly below the eyrie where more bones and pellets lay hidden in the tangle of grasses. These were added to the bags and the finally, after untying and coiling the rope, we checked for any more items around a plucking post used by the adult birds to sit on while preparing any items of prey for delivery to the chick. Once fledged, the chick could sit anywhere in the area as it had shown on our approach, so we made a meandering exit from the corrie and gathered a few more tit-bits of more recent prey.

Water voles are a not uncommon prey item at golden eagle eyries in the Highlands - a good indicator that they are still there, for elsewhere in the country, especially around lowland waterways, they are now very rare. This is probably a result of predation by feral American mink which feed on them, although there could be other reasons for their decline.

The size of eagle pellets varies depending on their contents. Small feathers or rabbit fur often come back up in a pellet the size of a Scots pine cone. One composed of large feathers or deer hair might be more the size and shape of a long potato. Grovelling around, picking up and dissecting all of these animal remains is quite off-putting at first, especially when they are wet and smelly or infested with maggots, but the information on the eagles' diets is well worth the discomfort. That visit produced the remains of a juvenile lesser black-backed gull, a herring gull, a red grouse and a raven's skull, not counting the pellet analysis. We went on to check another site in the area in the afternoon and found the remains of a short-eared owl, a red grouse, a juvenile cuckoo, and the lower hind limb of a large black-faced lamb. In addition, there were eight pellets that were mostly composed of various feathers.

In his book on the golden eagle, Jeff mentioned how he was impressed by the prey items found at one particular eagle's eyrie. Well, I was with him, as it was one of the nests visited on that trip to the west, and I can confirm that the prey list was impressive. The most that I have yet to see. The prey, not including that in pellets, was: four herring gulls, one fulmar, one hooded crow, two ravens, one red grouse, one peregrine, one kestrel, one merlin, one hedgehog, several rabbits, and two red deer calves.

The diversity of species that eagles have been feeding on reflects the availability of prey in their territories. In the north and eastern Highlands I would expect to collect large numbers of remains of mountain hare, red grouse and ptarmigan with varied local additions such as rabbits, and seasonal items such as deer calves and golden plover. The land there is open high hill ground and used mostly for grouse moor and deer forest. In the south-west, the history of the land's use has produced a matrix of open high ground, deer forest, sheep walk and enclosed land. And there is a long accessible coastline. Although golden eagles select to hunt ground dwelling mammals or birds of the size of rabbits and grouse when available, the feathers and hairs found in their eyries suggest that they are agile and versatile enough to catch almost any species around that size range. Lesser black-backed and herring gulls are regularly seen flying or loafing around the hilly peninsulas. They also regularly fly through the glens and passes to gain one coast from another. An eagle could catch one at either opportunity.

The raven is a close companion of the golden eagle, often sharing the same carrion - although not at the same time, the eagle dines first. The raven might be a competitor, but as the skull in that nest showed, it is not an equal, and it can even become prey. A raven being chased by an eagle must be a tremendous spectacle of cunning and flair, for the raven is a true master of flight and the eagle though large is anything but clumsy. Or, then again, less dramatically, the eagle might have surprised the raven while it was on the ground or at roost.

The skeleton of a meadow pipit lies below a golden eagle eyrie. There were no remains of rabbits, hares or grouse in the nest. Had the eagles switched to eating smaller prey than usual because their main prey were, for some reason, not abundant that year?

I can picture the capture of the short-eared owl more easily due to its lifestyle. Short-eared owls hunt open moors and wet flushes for voles, either by low repeated patrols of the ground or, more vulnerably, from a perch on a prominent rock, stump or post, where it would have been easily ambushed as it sat with its head down listening for voles in the grass. Or it could have been a female owl taken off the nest which is always set on the ground in the moor. Any owlets would equally be susceptible, as would any small bird in the moor.

While studying merlin breeding in Lewis, I found that eagles were their main predator and they ate both adults and young. All the merlin nests were on the ground and some were poorly concealed by grasses and straggly heather. That was to my eye near ground level. To an eagle's eye up high above, the young white downy chicks would have been obvious and easy defenceless prey. The adults were females and possibly the eagles noticed the birds flying into their nests and followed with a killing stoop.

Another ground nesting raptor which is taken by golden eagles is the hen harrier, which hunts over moorland, and from my own experiences if there are eagles nesting, there are unlikely to be hen harriers nearby. It is debatable whether they deliberately kill these potential competitors for the same prey (harriers will kill grouse and young hares as well as their main prey voles) or simply find them easy prey because they are more visible than secretive grouse or hares. Whatever the reason, any animal that is smaller than an eagle is potential prey, even if is a predator itself.

Cuckoo chicks are frequently taken and I have seen the entire nest of a meadow pipit in an eagle's eyrie along with the feathers of a large cuckoo chick. The eagle had obviously grabbed the whole nest and contents when it had dived onto its target and carried the lot back to its chicks. I once found an entire chaffinch's nest that had likewise been pilfered – evidence of an eagle hunting over open woodland.

The occurrence of the leg of the black-faced lamb at that first nest is not unusual; although most sheep remains that I have found have been in pellets during the winter at roosts. In some parts of the Highlands I have seen more than two dead sheep per square kilometre in late winter and such abundance of carrion actually maintains a higher density of eagles in winter than the live prey-rich eastern moorlands.

How eagles catch their variety of prey could be inexhaustibly debated and will take many more years of study to prove. What the golden eagles eat can be more easily determined by using the methods described above. Those details were of only two visits to eyries, over the years I have made hundreds of nest visits, and there seems to be a bewildering and ever increasing selection of prey species. I have seen the skull of a heron with its dagger beak, which given the chance could have done a great mischief to any eagle. Kestrels, both adult and nestlings might be fierce little falcons, but if an eagle can catch and kill the even more agile merlin or empty the nest of a brood of peregrines, then the kestrel becomes relatively easy meat. The white flight feathers of common and black-headed gulls are usually easy to spot in an eyrie and likewise these birds are probably easily picked out by eagles as they fly through the glens, or rest on the shores of hill lochans. Fledgling hoodies as well as adults are commonly taken in the west and the isles. And from the ducks, there are adult mallard and teal, the two most common ducks in the Highlands, along with the half-grown ducklings of red-breasted mergansers. Whether the ducks were taken in the air or on water or land I cannot say, but that does not hold for the ducklings, I have never seen young mergansers stray farther than the shoreline from water. They dive expertly when alarmed and can swim fast. Yet one eagle had killed two, likely from one brood, having probably returned to the same brood to catch more after catching the first. How did it do it?

Deer calves' remains, of both red and roe, are common in eyries, although only for a very short period, about two weeks. That is when the calves are new-born and lie in the heather and grass while their mothers go off to graze elsewhere. Red deer calves are more vulnerable as their speckled coat which gives good camouflage in woodland is much less effective on the open hill where most red deer are born. Once the calves are at foot, they are seldom at risk as their hinds and does protect them.

From the list of other mammals' remains found in eyries I can cite fox, badger, wildcat, domestic cat and otter, all of which would have given a good fight for their lives I think. Although the parts of foxes and otters I have found were from young animals. Other interesting corpses have been of red squirrels, perhaps dramatically plucked from a lofty bow of a pine. And decapitated voles can appear on the edge of eyries during vole plagues, laid out in a row just as they are on the edge of an owl's nest.

My most fascinating surprise among the mammals so far has been that of hedgehogs. When I first saw the unmistakable spines in a nest I was amazed. Not only that an eagle should chose to eat a hedgehog or how it opened up the spiny ball, but how and where the bird had found the hedgehog in the Highlands. Do these mainly nocturnal animals frequent the open hill during daylight? Or did the eagle hunt low on the fringe of woodland and inby land far down the glen at dawn or dusk? I have now seen remains of numerous hedge-hogs at several eyries, they are evidently a common prey item when and where available.

Throughout the golden eagles' range in Scotland I have been repeatedly astounded by the appearance of the immediately diagnostic, pungent odour and dusty grey feathers of the fulmar at far inland eyries. The fulmar only bred on the far out-lying islands of the St Kilda group until the 1890's. Since then they have increased their range around all the coastline of Scotland. Nowadays, they are so common a prey species that some individual eagles that nest close to nesting fulmars could be said to be specialist feeders on fulmars. In the Hebrides fulmars are frequently seen cruising through the glens, even in the western and northern mainland they are a not uncommon sight. But, how are they caught by the eagles whose home ranges are well inland? Perhaps they are more common visitors to the central hills than we think. I once encountered a fulmar flying past me in the heart of the Cairngorms, a full one hundred and twenty kilometres from the sea, about as far as one could be in Scotland. Do the eagles catch these odd wandering birds in flight or do the fulmars land occasionally and become easy targets due to their awkwardness on the ground? The latter is less likely considering that fulmars only tend to alight at their breeding colonies, most of which are on sea cliffs. Fulmars maybe use the Highland glens as short cuts between the North Sea and the Atlantic and are struck down in flight. They certainly fly in a steady glide that would be easy for an eagle to focus on. However eagles take fulmars, it is another example of how versatile they are at catching a variety of prey. Although in this case it might not be such a good thing.

The fulmar is a recent alternative to the golden eagles diet in Scotland, and these pelagic scavengers are known to hold high concentrations of poisonous chemicals in their bodies. These in turn could build up in the eagles and ultimately poison them. Eagles on the west coast have higher levels of PCBs (polychlorinated biphenyls) and mercury in their bodies than those in the eastern hills. What at first appears to be a saviour to individual eagles with scant food reserves in their home ranges, might cause losses in the long-term.

Scotland is a wonderful place to study golden eagles. There is so much variety of landscape, geology, soils, plants and animals in such a small country. And the human land use over the centuries has effected and continues to effect changes in the prey available to eagles. They are a very adaptable species, but they need space.

The eagle nest where Jeff and I found the long list of prey items is one that is no longer used. In the interim, the cliff across a gully from the nest has been developed for rock climbing. The rock is gabbro and very good to climb on, I know, I am a rock climber too and I enjoyed the freedom of safe movement on the wonderful holds when checking the eagles' nest. The climbers who discovered the cliff probably did not realise that the eagles nested there, however the damage is done now and another precious eagle nest site is lost to their use.

A well-grown eaglet, about ten weeks old, sits in its eyrie. Most eyries on cliffs are set on vegetated rock that is unattractive to rock-climbers. However, there are some that are on cliffs where people do climb, and many of these have been deserted, presumably due to disturbance by climbers

18

FIRST FLIGHT

Golden eagle nestlings take over ten weeks to fledge. By then the young birds are as large and impressive as their parents, their feathers are predominantly a dark chocolate colour and their wings hang heavily with the rich blood that fills them while their quills gain their full length.

Five kilometres into the moor there was a sudden shrill tchooking call from a greenshank as it flew from the edge of a lochan. Immediately, its mate joined in, and the pair flew frantically around me, stooping and soaring, then landed on a hillock of moss where they piped their long bills for all they were worth. They obviously had chicks which had been feeding, and were now hiding, in the sedge around the rim of the pool, so I spurred myself on to alleviate their alarm and to escape from their deafening din.

At the far end of the same boggy hollow a mouse flitted from my foot and ran squeaking and twisting through the grasses. Then it opened a pair of wings and lifted its head to turn and look for my reaction, which was to stop. I wasn't going to be fooled by the rodent run distraction display of the dunlin. Unlike a fox, or any domestic dog, I had seen this behaviour before and have learned from experience that there must have been a nest or chicks close by. Without moving, for fear of stepping on the eggs or young, I scrutinised the hollows in the tussocks around my feet and slowly defined the entwined bodies and bills of four delicate newly hatched chicks in the nest about two metres away.

Altogether, the brood weighed less than an ounce. Their camouflaged down was a perfect dark peaty brown, dotted with grassy yellows and fawns. Entrusting their lives to these colours, all four heads lay still on the warm downy backs of their siblings. They were obviously newly hatched and their egg teeth were still sharp and shone pearly white at the point of each pencil tip bill. While I stood by the nest, both parents ran around me, trilling like clockwork toys less than three metres away. Now I knew where the nest was I could step on confidently and not trample its delicate contents.

I never saw the greenshank chicks, they would have been older, less downy, and well scattered in separate hiding places. The breeding season was well on; it was a surprise to see the dunlin hatching. The eggs must have been laid late, perhaps as a re-lay after losing a first clutch to a crow or fox. As it was late in the breeding season and most young birds had fledged, the rest of my walk was largely uneventful. My feet became lighter after those two encounters, long walks over moors are much easier when there is something to see. It was a brief lightness however; life on the moor was so sparse. Further on I flushed a red grouse from a heathery bank. Then I passed a dead sheep. All the flesh had been well picked from its bones that lay scattered around with tufts of fleece. Every scrap of food had been cleared.

As I gained the top of a broad ridge I stopped to scan over the other half of the moor. It spread out on three sides with the ridge on the right sinking into a sea of flat peatland. On my left, a mile long cliff climbed higher and higher as it stretched into the west. Tall columns, pillars, and towers, all reared up between deep and horribly wet gullies. The sandstone was crimson where it was wet and pink where it was dry. Not much was pink.

I lay myself down on a fine table-sized slab, my first dry seat all day. The cliff grew bigger still as I spied through the telescope and meandered from ledge to ledge. Then, there it was, that year's eyrie. Despite its dwarfed proportions, it was easily picked out as a faint trail of whitewash and down scattered for two hundred metres below. The birds had chosen a dry ledge, tucked beneath a square cut overlap at the top of a buttressed pillar. A single twisted rowan gripped onto the lip of the roof. Its grey gnarled limbs trailed down, it was alive, full of leaf and I wondered how long it had been since the seed was dropped there on that Gormenghast crag.

At the back of the nest the burnt ochre brown shape of a very large eaglet was bouncing on the nest. Then it flipped open its wings revealing the white down on the underside where the last of its feathers had to grow out of their quills. The young bird folded and stretched its wings again and again, and it jumped out onto the front edge of the eyrie, right onto the very brink. It seemed frustrated with sitting alone on its nest for the past two months or more. From the eyrie, it would have been able to watch everything below its domineering cliff. How often had its eyes traced the streams' twisting lines from the lochs to the sea? The sea, that vast open space, deceptively defined by the arc of the horizon. Like a bird in a cage, the eaglet kept jumping back and forth from one side of its eyrie to the other.

Suddenly, it gave an extra strong heave and flapped onto the front edge as if about to launch off. It tottered and clung deep into the sticks and heather with its talons. It toppled forward and swung down over the edge, scrambling with half-folded wings and one foot while the other gripped the nest. A twist, a flap of wings, and it clambered back onto the nest. When it had re-settled itself safe and secure, it began to call loudly in a shrill anxious voice. The chick was excited. It flapped and it jumped and it yelped. It almost flew, but had held back just at the last moment. I took my eye from the telescope and immediately saw the dark form of a parent bird glide over from behind me. Its talons hung low below its tail gripping tightly to the body of a fulmar. The prey could be clearly seen through my binoculars and easily identified as the eagle passed right over me. The eagle flew directly to the eyrie without a beat of a wing and then, tormenting the nestling, continued past it in the same solid glide to land on a heathery slope on the top of a neighbouring buttress about a hundred metres beyond and below. Once more, the scale of the cliff was brought to my attention as such a large bird overhead became lost on the vast wall.

The chick became wild with excitement, or was it just hunger. It yelped like a puppy calling for attention from its mother. It was given nothing more than answering yelps. And then as if to add further torment the parent proceeded to pull at the fulmar and toss tufts of white feathers away on the wind. The chick clung to the very edge of the eyrie, stepped out onto the air, squawked and pulled back with a deep reverse thrust of its wings. Having prepared the meal, the adult left the prey and glided onto a nearby ledge where it began to pick and preen its talons and feathers. Room service had been terminated. The eagle had flown out to the sea cliffs, snatched a fulmar from the cliff larder and brought it back home. And that was it, no more bedroom meals, the youngster would have to come down to dinner.

The chick had probably seen and recognised its parent long before I had, and that could have been why it had been acting so excitedly when I first saw it. Anxiety oozed from the chick. Its eyes were now intently focused on the food. Its neck strained as it leaned out over the edge as if to check the drop. Its talons were clenched. And its wings flapped harder and harder. I could see that its wings were actually lifting it, the chick wasn't jumping, it was flying up and down. Then, maybe with intention, certainly not with full confidence, the eaglet finally let go of the eyrie and dropped down onto the wind on wobbling wings. It manoeuvred a course, more with its wings than its tail as an adult bird would. None the less it managed to land on target close by the fulmar. Well, perhaps about fifteen metres away, which wasn't bad for a first effort. The large open slope offered a fine soft crash-landing site for half-folded wings, and the tight grasping talons clung fast to the heather.

It was done. The eaglet had left its eyrie on its maiden flight. I could hear a corny soundtrack deep in the back of my mind – the sugary voice over and triumphant music. I snuffed it out quickly and promised myself not to watch any more of such films. It disnae happen like that in real life. The chick was lying in an unceremonious heap, panting, bill wide, wings drooping, eyes wide. And it lay like that for several minutes while it seemed to assess the situation it had got itself into. Then it stood erect, gave itself a quick shake to re-orientate after the shock of the big wide world, strode up to the prey and greedily cast clouds of white feathers across the cliff face.

The Falls of Tarf in autumn spate.

Young eagles disperse in autumn which, with shortening daylight and many wet rainy days, must prove a testing time for the birds.

This Scots pine has held a golden eagle eyrie for more than a hundred years and was used frequently until the turn of this century. Birds still touch up the nest, but the recent increase in the number of hillwalkers and campers passing close by seems to have become intolerable to the eagles.

The tree would have been mature when the birds began nesting in it and there is barely any change in its size or shape over the years, only the canopy is thicker these days - see *The Golden Eagle, King of Birds* by Seton Gordon 1955. The dead tree lying at its feet was already long fallen when Seton Gordon knew it. It was probably felled as many were in the glen on the night of 17th November 1893, by the same storm that felled thousands of trees in the Highlands. So how old are the trees in this and other woods in the Highlands? How long will it take for young trees to replace them, and become strong enough limbed to bear such a heavy eyrie?

19

CALEDONIAN WOOD

The remnants of the old Highland pinewoods are a national treasure,
the large eagle eyries in them are natural wonders of architecture.

As I entered the wood a soft calm drew around my shoulders. Sheltered from the wind, I straightened my back, relaxed, and walked on.

I was in a stand of tall individually shaped firs - the old local name for Scots pines, and I was in an old Caledonian wood - a modern name derived from Caledonius Saltus which is inscribed on an illustration of a large wooded area in the central Highlands on a map of Scotland drawn by Ptolemy circa 140 C.E.. The Latin word saltus translates as a wood with glades, which suggests that even then the land was not continuously tree-clad. Caledonius is stems from the name Caledonii, the local tribe of the area. Although nowadays the terms Caledonian wood or forest are generally used in reference to the remnants of the once extensive pine woodland; the birch, juniper, rowan, aspen, holly and alder are all just as much part of these ancient woods.

Each fir had a character of its own. Some of the well spaced trees were thick stemmed with long drooping branches, others were multi-stemmed, and those in close groups had tall straight trunks topped with tight bushy crowns. At the other extreme there were tiny bonsai pines growing out of clefts by a waterfall, in amongst gardens and curtains of moss. One slender pine stood all alone, set on a mid-river island encircled by steep angled rock. It was erect and well formed though stunted, growing as it was on a thin layer of soil on top of rock. The overcast light saturated the reds of the bark and the greens of the mosses. It was rich woodland; twitters and chips told me there were siskins and crossbills up in the canopy, and squirrels had dropped chewed-up cones on the floor.

My world was small though. Over on the far bank, a deer fence had been strung around newly planted pines, set in monotonous rows, amid the skeletal limbs of their felled elders.

Farther uphill, the pines opened out and blended with light airy birches. The tussocks of heather and blaeberry grew thicker there than under the canopy, and deer had nipped off many of the succulent shoots from both plants. No deer were about, but a cock capercaillie was nibbling the buds from a crown of a pine up above; he also knew the nutritional value of shoots. I am familiar with capercaillie, yet every one I come across re-impresses me with its size. He was huge, and I was incredulous, how could such a large turkey-sized bird perch on a thin treetop branch that bent right over beneath him. On seeing me, he froze - for a second, and turned his eye from the sun that shone behind me. One more step towards him confirmed me as a possible predator and he was off. He caught his weight on his big black wings and rushed off, part under, part through the branches of a thicket. Although out of sight, I could hear his wings whirring and his bulk smashing its way through, then he landed in a tree not two hundred metres away.

As I climbed up the open wooded hillside, an increasing number of grey trunks held up their withered limbs. Stripped of their bark, the dead pines showed their twists of spiral growth, natural spring, that helps these amazingly strong trees stand up to the fierce winter storms that whirl down the glen. Even in death, Caledonian firs are magnificent. So much so that I don't mind to see them die, that is nature, except that I could not see any sign of the next generation and that is not the natural way. Tree seedlings don't live for long in most of the Highlands. As soon as they pop their heads up above the level of the heather, or more especially the snow in winter, they are chewed off by the red deer, which roam in herds that are too large to give the woodland a chance to maintain, or rather regain, its structure. Sure, there are thousands of pines spreading over the hillsides in plantations, but they are of a different seed; they don't continue the native lineage. Nor do they have the same variety of growth forms, they are mere telegraph pole clones. There is much more to a wood than trees.

Birches stood around suffering the same problem, all old and gnarled. No young trees could be seen amongst them either. Great-spotted woodpeckers had been drilling into the decaying limbs, casting splinters and sawdust onto the heather. They were picking the bones of the dead in search of grubs in the soft rotting tissue. An elliptical hole near the top of one stump showed where a pair had reared a brood in the past. The local bird that had left the new woodchips nearby might even have still been using the hole as a roost site. In there it would be tucked well out of any cold wind. And below an old fir's split trunk nearby, I found pellets cast by a tawny owl. It too had found a roost in the shelter of the wood.

I climbed out onto the moor above the wood. The ground was wet, and in between the black woody stems of heather a medley of mosses and grasses held water in a neat trick of suspension, releasing it all onto my legs as I waded through the deep tangle. And when I broke my stride with an occasional trip, I sank my arms into the ooze and they were soaked too.

I staggered towards a finger of wood that pokes out over the moor where a shallow depression gives a pocket of shelter from the winds. There are some big trees there, leaning away from the prevailing westerlies. Smaller individuals lie scattered around the edge of the moor, perfectly formed but impossible to guess of age. Their small girth and tight crowns have probably taken as long to grow as the tall firs in the wood down the hill. The trees on the windward edge of the grove were also slightly restrained in height and offered tight clusters of branches and needles in defence to the gales. The tallest of the group stood on the leeward side, their long strong branches spread open and wide. The limbs were far enough apart for an eagle to fly through, they were sturdy enough to support the weight of a golden eagle eyrie, and in one of those trees there is a large old nest. I have known it for more than thirty years, and it was there long before that.

A collection of mountain hares' big hind paws lay around the base of the nest tree. They are all that are usually left of a hare by an eagle. Perhaps they find the large heavily furred paws too awkward to swallow - the stiff bushy fur gives them a snow shoe effect which helps the hares run across the winter hills easily. A few grouse feathers also lay about, and that was all the prey remains on the ground. A half dozen or so pellets were likewise mostly composed of mountain hare fur, and a few grouse feathers.

The branches of the nest tree were damp and cold. And as the branches all slope down from the underside of the leaning trunk, my climb up to the eyrie was a careful one. I had climbed that tree many times before, but that didn't make the climb any easier, as the main danger was in slipping from the wet branches. I pulled up onto the first branch, balanced upright, teetering until I grabbed the next branch, and so on, repeat, repeat. The eyrie is twenty metres up. The nest is massive – two metres deep and much the same wide. When I reached it, I had to squeeze through branches to get past its huge bulk. Then I plumped myself down in the middle of it with ample room to turn around and rake about in the detritus for scraps of prey. The giant collection of fir branches and heather stems were entwined into a mass deeper and wider than, for comparison - one of those straw bales that lie in the fields after harvest. My extra weight was of little significance, for a foot of snow lying on the platform would be heavier than me, and that eyrie must have held that much snow many times over the years. I was confident, relaxed and comfortable.

The nest cup had lost all the shape and colour of a fresh nest with lots of fresh green pine twigs when first built. It was now just a big pile of grey sticks with a dusty grey layer of decayed fur, feathers and scales. The scales were from the cast feather bases and feet of the growing chicks. A pasting of guano clung around the edge and on the innermost branches which were well sheltered from rain. There were more hare paws and grouse feathers and lots of cast downy feathers from the fledglings. It looked as if both the fledglings I had seen in the nest earlier in the year had flown successfully. There was no evidence at all of any having died in the nest.

The descent was more awkward than the ascent as the branches nearest the nest were all green with slime growing on the nutrient–rich goo that is washed down from the nest. Lowering one's weight onto dodgy footholds is always more tricky and insecure than pulling up through what the eyes and hands can test first. As I was looking down all the time, I noticed a cast eagle's feather lying point into the heather below. It was spinning in the wind, slowly twisting itself down into the vegetation.

I dropped the last few metres from the lowest branches and picked up the feather. It was a secondary flight feather, cast by a female (it was larger with a thicker quill than one from a male). Then, still scanning the ground for feathers and pellets, I walked quickly away down the line of trees. And as I entered a clearing I saw the wings of an eagle dodge around the brow of the hill up ahead. One second, perhaps two, that was all that I had of a glimpse. It was an adult bird, and judging by its small size it was a male. That was it, my only view of an eagle all day. Yet, from a little detective work, I had built up a case, which would help to fill in the gap in my knowledge of that pair's breeding success that year.

I crossed over onto a rocky point which looks right out over the woods below, down the glen, up the neighbouring glen, and onto the hills beyond. It was a marvellous view, spanning from the cultivated inbye land in the lower part of the valley, over abandoned crofts and shielings farther upstream, and into the high corries beyond. Spread below me was a living land that had changed many times over the years. Old dry-stane dykes and long rectangular ridges of turf told where families had once cleared and set up a small hold on the land. White dots of sheep could be seen on a far hillside. And herds of deer lay up in the lee of the ridges. Strips of burnt heather had been burned to improve the moorland for grouse.

The old Caledonian wood, of which I was in a part, was probably at its greatest extent about 5,000 years ago. Since then it has retracted and divided into fragments during millennia of cooler climates. Large tracts of trees would have been cleared as the first pre-historic settlers moved in to the hills, tilling the better soils for crops and grazing cattle on the poor ones. There would have been continuous felling or burning of trees for centuries, just as there is ongoing in many countries today. The people would have been cutting trees for firewood and timber, and they needed crops and animals to live, just as those people in other parts of the world do today. Then, 150-300 years ago, logging of Scots firs became big industry; a great many trees were felled, and more have been felled since. As timber had become an industry there was financial value in replanting, and many have since been felled and replanted. However, while these plantation trees have been growing, the numbers of red deer have been increasing, and outside the plantations browsing by deer has largely prevented the establishment of native seedlings.

The youngest of the tall firs testify to this history. They are at least two hundred years old, an age slightly older than that of deer stalking as a fashionable Highland gentleman's sport, and the accompanying growth of the deer herds. I looked down at the changes and thought of all I had seen that day, and the plight of the Caledonian wood; the birches, rowans and others as well as the firs. It wasn't all in despair. There is now value in conservation and wildlife management; a new attitude is slowly turning towards culling red deer in realistic high numbers, enough to allow seedlings to stretch up and woodland to develop. The trees will be slow growing; I only hope they will be large enough to bear eyries before the current holders die.

I know of several eagle eyries which have been built in plantation trees (two have been felled for timber). Clearly, eagles can use these trees as well as the wild trees, but the lifespans of the nests are determined by the market value of the trees they are in. Also the nests are usually much smaller than those in the old firs, as the branches are thinner and more tightly set, allowing less scope for large eyries.

A view south from above the Lairig Ghru in the Cairngorms - a favourite flight path of thousands of pink-footed and greylag geese every autumn, as they head from Iceland to their Scottish lowland feeding grounds. Golden eagles are perfectly capable of striking down geese in flight, although I have never seen so myself.
Perhaps on another eagle day......

20

WHEN THE GEESE COME IN

The last dreamy days of summer are swept away by low storm-laden clouds. Rain stirs the burns into brown frothy spate. Wind knocks the leaves from the branches. Then the storm abates and gives way to a few more blue skies, revealing a white dusting of snow on the highest tops and an orange tint on the moors.

Under the clear skies of one late September night, the first true frost bit the air. It was a typical first frost, soft, and a heavy dew was all that was left after dawn. Sheets of droplets lay on the grasses, and diamante threads hung from cobwebs in the brambles. The first peep of sunlight brought the jewelled beads to life then destroyed them all at once with its radiant warmth. The clean air was silent. No, it wasn't, I could hear the calls of geese, far off geese, and pin-pointed them high and directly above. Long, jostling arms spread over two hundred birds behind the leader. Then another skein came over the Cairngorms, or rather, through the Lairig Ghru, the high pass that splits the Cairngorms in two. The skein was followed by another, then another, and another. And they were flying through two other passes that I could see, the Lairig an Laoigh and the Sneck. There were geese all over the sky and there was to be a continuous stream of geese passing over me for the rest of the day.

The geese had timed their migration well to catch the favourable gap in the weather for their journey. They were greylag geese, easily identified by their deep domestic goose honking which is less musical than the calls of the other common grey geese in the hill passes, the pink-footed geese. Although they jostled for position within each chevron, the geese as a whole kept a steady north to south course. North from the breeding grounds of Iceland, south to the arable lands of Tayside and Fife, or farther south and into England. Those birds would have left their breeding grounds the previous day, flown through the night and now they had almost reached their wintering grounds on this the next day, in one long flight.

How many times had the leading birds made that passage? Did they know the hill passes and aim for the routes they knew best? Had we seen each other in the hills before? Over the years, I felt that we must have. I marvelled at the way those geese edged around the ridges and deftly slipped though the passes without wasting any energy gaining more height than was necessary. A bit like myself really.

The hillsides on either side of the pass, Ben Macdui and its satellites to the east and Braeriach and its neighbours in the west, were iced with fresh snow and the lower summits around them were sugared with frost. I too had timed my journey well. I was out for a quick walk around a few of the lower Cairngorms peaks before the next forecast weather front came in that afternoon. Once I was up on the top of the hills the walking became easy. The dwarf heather there is either stunted by the wind or it lies prostrate behind sheltering rocks. Peaty soil lying in a bealach was frozen hard and beared my weight as I crossed. It is so much easier to walk on peat in frosty conditions than in summer when the ground is soft, wet and tedious to plod through.

In the glen down below, other peaty hollows were still soft and wet. And I had noticed on the way up that many had been rolled in by red deer stags which were well into the rut. The rut, or mating season, is brought on by the shortening daylight of autumn. The stags wallow in mud and peat, smearing their coats black to give them bold heavy outlines. This is further enhanced by a seasonal growth of a shaggy mane on their throat and topped by their wide spreading antlers, which are themselves enhanced with peat and heather by rubbing them in the wallows. The whole is adapted to give as fearful an impression of size to other stags as possible.

The stags' challenging roars to one another echoed across the corrie below me. Forced starts of short grunts were followed by long drawn-out roars, billows of steamy breath oozed into the cold air from outstretched mouths. There were stags running about herding hinds into groups, chasing away young or weaker suitors. Others paraded their might by strutting back and forth broadside to rivals, showing the best of their bulk. One charged head down at another. The two then battled and grappled with entwined antlers, knocking and thumping with the full weight of both beasts behind every blow. Their rattles resounded far off; their grunting and snorting was smothered in a hoof-cast spray of peat. Then the aggressor routed the other's rear as he retreated. Such battles are actually rather rare. The main plan is to gain as much weight before the rut as possible and grow the largest antlers. Then strut around and pretend you are tough. This usually works, and the largest and blackest stags can have forty or more hinds held in a tight harem. But of course, any stag with that many hinds is kept busy defending them from others' attentions. So, cheeky young or small stags often sneak in to cover the hinds while the sultans are busy fighting amongst themselves.

The whole corrie had the air of a battleground. There were the war cries, the flags of antlers, the blackened wallow holes, the pungent smell of stags exuded over the ground where bodies had rolled, and the rich taste of victory completed the image as a master stag roared and rattled around his prize parcel of hinds.

Out beyond the corrie, the scene was more peaceful. There was a scattering of stags wandering alone in pursuit of hinds, or lying exhausted, for stags do not eat much during the rut. And their distant roars came across only as faint wails. The wind was picking up from the west, and the majority of the deer were being herded by the breeze onto the sheltered eastern slopes which I had just passed. I looked out over the hills and moors and conceded that summer had finally given way to autumn. The blaeberry leaves had turned to dry rust and the berries were now grey, wrinkled, and well past their best. The purple bloom had faded from the heather. And in contrast with the sombre landscape, a previously reticent rowan tucked into a bend of the burn, now stood out loud with its bright red and yellow flash of berries and leaves. Then a young golden eagle rose out of the canvas.

Soaring in a chain-linked pattern, it slowly traversed the glen. It was a dark bird with contrasting white wing and tail patches. There were no sun-bleached or cast feathers; all were a fresh new-grown cocoa colour and the set was complete. It was obviously a young bird, so I expected to see one or two attendant adults, for the youngster could only have left the nest a few months previously and I have seen young eagles close to their parents as late as November. The area which that bird was hunting over was usually hunted by birds from a nest site in the next glen, so I spied in that direction. I couldn't see any other birds. It might have been out on an unescorted sortie, gaining confidence to explore the world on its own. I'll never know, as the bird was unmarked so I didn't know where it came from. That is exactly the type of information that is being collated on eagle behaviour from satellite-tracking results – how the youngsters begin to move away from their natal site. How far do they go? How long do they go for? The eagle kept turning and looping its way over the moors, its dull colouring fading into the dull background, and I lost it.

I have been up on those hills at the same time of year before, numerous times. The geese aren't always migrating past on the same date, that depends on the weather, but on one other occasion when they were, there were three eagles up in the sky. Although they were too far away to make out their plumage (they were mere specks in the blue), I assumed that they were the local pair and a young bird of the year. There are reports of golden eagles striking down migrating geese as they fly through mountain passes and I was keen for the chance to see them do so. Those birds on that day were certainly interested. They soared above the geese, all three in the same circuit. My eyes were glued to the binoculars, terrified of losing the birds or missing the action, when suddenly, one took a steep dive at the geese. I thought this was it, a kill in mid-air, but no, it pulled up short. The same bird made several more dives at the geese in as many minutes and then all three swung away out of my sight. Had it been a half-hearted attempt as the birds were not hungry, or had it perhaps been the youngster practising? All I know is that the eagles never actually killed a goose in flight. It was a marvellous flight display none the less. Perhaps I'll witness a strike another day.

Those three eagles disappeared as stealthily as the young one I had just lost. And as perfectly timed as ever they do, that bird had disappeared as a bank of clouds smothered the area. The eagle had read the weather well ahead of me. The fine clear skies of the morning were now screened by a mist that fell lower and lower until it lay half way down the sides of the glens. I turned about and made my way back, following my mental map of the land.

Geese had continued to come through the hills the whole day, and now that I was blinded by the mist, I could hear their calls louder than ever. They had been forced down by the thick cloud, and the hardy travellers were now skimming low along the glens beneath the ceiling. They seemed to be following the hillsides that lead south. With the mist swirling around my ankles, I stopped and listened. A skein came within fifty metres of me on my level. Their honking was faint to begin with, and the whistling of the driving mist smothered my ears. Then as their calls reached a cacophonous clamour I could even hear their wingbeats, and for a second or two, that was all. I could see grey birds in a grey mist, with black glass bead eyes and orange bills. The cloud was falling quickly and I counted a line of fifty or so before the line beat on into the fog, and the haunting wild encounter was over.

I carried on my way and soon the mist had turned to rain, real rain, great big plops of it. It trickled down my spine, I shivered and pulled up my hood. I was reluctant to do so as I don't like wearing a hood, they shut out too much sound, and so much of a day in the hills is absorbed through the ears. The stags were still roaring all around me, out of sight, and now with my hood up, out of earshot. The sight and sound of the geese and deer in the clear morning had held the crisp clean aspect of the last days of summer. Then in the afternoon, the wet dull glimpses of these creatures epitomised the darker aspect of the approaching autumn. And as I drifted down the glen, isolated in my little weather-proof bubble, I thought about another time when I had seen autumn come in on wings.

Again, I was in the Cairngorms. The sky was high but hazy, and I was rounding the shoulder of a hill at the south end of a pass. About fifty metres ahead three long white dashes came around the corner and bore down on me. Within the split time that it took me to recognise them as whooper swans, their huge deep pulsing wingbeats were throbbing over my head. They rose to avoid me – that's how low they were to the hillside, each cocked an eye and once past, they dipped down to their initial flight level. All I could do was sit down and watch them, no not just watch them, but wonder at them. They were magnificent and what a setting, deep in the mountains, so far from where I would expect to see whoopers. The trio were a family party of two full adults with pale yellow bills and bright ivory plumage accompanied by a dirty white juvenile with a grey bill - a single offspring of that year. Like the geese, the whoopers had migrated to Scotland from their breeding grounds in Iceland. I lay back in the heather and stared after them. The throbbing faded from my ears and their white dissolved into the haze of the hills. Autumn had arrived.

Our senses are roused in autumn with the colours of leaves, the sounds of birds, the touch of frost, the smell of moist leaf litter, and the taste of berries.

21

NOVEMBER DAYS

Short dark days can be surprisingly good for seeing wildlife as all the diurnal animals, including eagles, spend a high proportion of these days active.

Cool damp air wicked the warmth from my body. The sodden branches of pine and birch drooped in silence. And an earthy odour of decay hung in the air. With both hands thrust deep in my jacket pockets I tramped on along the dark woodland track leading out to the glen. Dim as the scene was, I was glad of the shelter provided by the trees, as the morning's light drizzle soon strengthened to driving rain. I quickened my step and dropped my head.

Everything felt wet, miserable and nasty, although in truth never really horrible. It was still great to be out. The puddles I splashed through were sprinkled with the colourful cast leaves of rowan and birch. Fresh flaky pine bark gleamed bright all through the wood and rainwater had added flashing white highlights all through the scene.

Up in the canopy, thin squeaky calls kept goldcrests in contact with one another as they peered and probed for the smallest of insects hiding amongst the needles. These, the most frail of Scottish birds, were all that dared venture out while the rain poured down. No other creature appeared to stir. Being so small with a fast metabolic rate, goldcrests must hunt for food to keep up their energy resources to help see them through the lengthening nights.

Gradually, as the rain eased and I walked on up the riverside, other birds, one by one, began to reveal themselves as they began to search for food. A momentary glimpse of white exposed a group of bullfinches feeding surreptitiously in a rowan tree. Bright red waxy berries bent the naked grey branches under their weight, and the birds hung down from the very ends of the branches, stretching their necks to pick their fill of fruit. Contorted in their exertions, white rumps, black caps and tails, the pink buff of the females' breasts and the red of the males' were all so complimentary to the cloud of scarlet berries. Floating whistles kept each in contact, although the soft notes drifted very faintly above the rattling of the burn. Not like the harsh zipp-zipp of a dipper calling as it flew upstream between us. Dashing black and white as it skimmed over the water, wings whirling away.

When I reached the top edge of the pines I stopped for a rethink. Beyond the comforting arms of the last few trees there was a bleak, dour aspect. The wet grey glen lay exposed to the cold sleety wind. There was no sense in going any farther. Even though the clouds were breaking up the poor visibility made the whole venture pointless. I would never be able to see any eagles in that mirk, so I chose to idle back along the forest edge and make the best of the day.

Although the weather was still wild on the high ground, the rain soon stopped altogether in the glen. And birds came to life all around. A troop of coal tits stitched the frayed ends of the pine branches. Ranks of redwings rolled through the rowans, squeaking and rattling all the way. They snatched up the berries, bulged their crops, and they cast their red sticky droppings all through the wood. Where had they all been hiding?

A herd of beef cattle on the fringe of the wood rummaged through their hay, steam rising steadily from their shaggy winter coats. A shifty gang of crows slipped away from their hooves as I approached. The churned-up and scattered fodder held rich pickings for wily opportunists. Up above, a flock of fieldfares bounced and chuckled, flicking their wings on toward the next rowan grove. I followed their line until they dropped into some laden branches, then as I lifted my gaze up to the skyline, I picked out the determined glide of an eagle swooping across the hillside. My eye caught it just in time as it left the blue grey sky to enter the backcloth of the purple-grey hillside – it's always best to watch the skyline for eagles as it is difficult to spot them against the hillside, where I needed binoculars to follow that bird. There was very little detail to identify the bird. But no plumage was necessary. A large dark outstretched pair of wings with a glint of low afternoon sun on the golden nape and shoulders were all that was needed to colour the characteristic flight.

Thirty seconds later the bird was gone. A single curving swoop, a wing flick here and wing flick there, it crossed over a ridge, and that was it, show over. Perhaps the bird was setting off on a late hunt, postponed by the heavy shower.

Or like me, it knew that there would be a burst of activity after the shower and potential prey would be out and about. The eagle had cruised out of a high corrie where it had probably been sheltering, and swept over the moor where red grouse would be moving through the heather. But, I'll never know the true purpose of that eagle's glide as it never came back into view. That was my eagle for that day.

The crows were now raking for food on the same ridge. They hopped through the stumps and branches of windblown trees and dropped into the heather to forage. I wondered what they were finding so much of. Despite the warm colours cast in the thin light, there was very little heat to beat off the continuing chill. I spurred myself on to build up some warmth, stepping high through the cattle-churned muck, and jumping the worst of the gutters and dubs.

I moved on to spend the last hour of daylight down by the loch, and by the time I got there one group of greylag geese were already settled down to roost on the centre of the water - a string of black buoys bobbing on the backlit silver surface. Common gulls poured in over the treetops from all directions in which they had spent their days picking at worms, leatherjackets or some other grubs wherever the farmers had been turning the soil after the year's harvest. The gulls slipped silently into the scene, twenty or so in one flight, a group of three in another. Then more and more. Once on the water however, they clamoured and jiggled for position. Their increasing numbers steadily weaving a grey plaid, row by row in the lee of the flock.

In contrast to the discipline of the gulls, a jumbled procession of redwing and fieldfares hopped, flitted and chattered through the scrub's bracken, heather and birch. Seemingly late in arriving that year, these Scandinavian winter visitors were now in their full force of marauding hordes, sweeping through the country in their ceaseless plunder of berries.

A buzzard swooped low through the branches, disturbed from its roost by the commotion. In turn it set panic amongst the flock, scattering the thrushes wide and low into cover, stifling their squeals – if only for a second or two. As the mob dived into cover they exposed a threesome of long-tailed tits which kept feeding without heed, tumbling from limb to limb of the birches.

Out on the water, the waves smoothed themselves out, and I settled quietly between a stump and a willow on the gently sloping bank so I could spy through my binoculars to see what was about. Whistling widgeon rippled a wake over by the far shore near a phragmites reedbed. Their delicate call fitted well in the then tranquil atmosphere. Then there was a rough squawking from the mallards hidden in the sedges. That was offsetting, yet in their own way evocative, as their calls echoed through the now frosty air.

Only the golden tips of the alder and willow which stood above the reed heads now caught any light. And there on their branches, sitting as quietly as shadows, were two female hen harriers. A few minute later another one swung low over the reeds from my left to alight in the same sprawling willow. All three sat with never a twitch, watching, just watching, as the pale light faded from the skyline.

Then again, from the left, a most handsome chalk-blue male harrier grabbed my attention – 'here comes some action!' He came in low over the open water, in a jaunty flight. He was still hungry, or perhaps he was simply an opportunist, for he patrolled back and forth along the length of the reedbed. And his persistence was rewarded. A skip, a twist, then down, he had caught something, a reed bunting most likely as there were hundreds of them settling to roost in the reeds. One of the females had been watching him too. As soon as he perched with his prey she flew straight towards him and rushed to snatch his dinner. He was having none of that and held her off with a flash of wings and wide eyes and bill. Once she was off he lifted up and shifted farther back into a thicket to disappear into the gloom. A pale bird in pale light, a ghost in the greyness.

Soon after, in the very last of the light, a fifth hen harrier darted along the full length of the loch. It was mobbed by a twist of three reed buntings, the last to settle into their roost. In the near blackness, her sexually dimorphic colour of dark chocolate brown was difficult to see, and I had to stare hard through the binoculars to follow her white rump as it dashed and dotted a path directly to the communal roost where the others were by then well settled.

She wasn't the last to retire. Skeins of greylags were building up to a climax. They gaggled and honked as they approached in a series of squadrons. Then as one, they went silent and dropped, in wobbly lines onto the water. Once on the water they continued their gaggle, as they preened and jostled into their sleeping positions.

I left them as faint shadows on the whispering loch. Those were the last sounds of the day; the whirring of air through their outspread primary feathers as they slowed down in approach right over my head, and their dwindling peeps as they tucked their heads into pillowed down for the night.

The loch lay still and the day was done.

A neck stretched up into the morning's twilight, then slowly curled back down. Far out in the centre of the loch, wings were spread then folded again. Honking and grunting, the geese were beginning to stir. Greylags are the noisiest of our geese at any time and never completely still during the night. At dawn they create a tremendous gaggle. Finally the whole loch heaved and yawned. Squadrons of geese took off one after another, as if none wanted to be last out. The air was full of wings, slow strong thrusting wings, beating twisting courses up and around, then out over the tops of the lochside pines.

It was quiet again. I noticed the white frost that nipped the pines' needles, the naked twigs of the birches and the curled dry leaves of the sleeping bog myrtle. It hadn't looked that cold when I walked to the loch in the dark. There was a change in the weather. Crunchy sponges of moss carpeted the shore where the water had shrunk back behind the reed bed, and ice edged on for another five metres or so out beyond.

Odd patches of water were still free by the shore. Such as one spot where the burn flowed in, and up the creeks in the reed bed where the mallard were dabbling and poking in the flotsam for food. Beside the mallard and scattered in ones and twos right across to the farthest shore were goldeneye. They were all diving to roam along the lochs shallow floor then popping up farther on. Soon the loch would be frozen over completely as a cold spell slipped in. Until then, the ducks had to feed hard to put on fat. For, when the ice came they might need to go far to search out another food source. The geese too would probably move too. Although, they graze in the fields, they use the loch as a secure roost. Once it was frozen over, foxes could walk across to them and catch them as they slept. The same would apply to the gulls.

The numbers of birds in the air increased with the rising light. Gulls radiated out in all directions, a pair of carrion crows slipped by on whispering wings, higher up a long flock of jackdaws chattered and cackled, followed by a rattle of rooks. And pheasants creaked and scratched as they edged out of the woods to feed on the stubbles. The handsome cocks' faces flamed red in the winter light. Equally vocal, though hidden deep out of sight, a water rail squealed as it set off on a day of stealth in the thick of the reeds.

Just beyond the edge of the ice, a concentric sequence of waves vibrated across the flat gold water. At the epicentre, a head bobbed up, then dipped below the shelf of ice. It came back up with a hump, as it curled its tail. Then there were two, a second otter was chasing the first. In splashes and tumbles, they reared themselves up till only their hindquarters and tails were left in the water. Face to face, their jaws open in play, their forepaws wrestled in pushes and pulls. Sinking back below water with barely a bubble, they swam in a chase for several metres. Together again, they twisted and rolled in a coil of bodies and tails. They eventually chased one another around a corner, in behind a curtain of reeds and out of my sight. One otter reappeared later, fishing the same spot beside a watchful mallard. It wasn't long in catching a small fish and promptly ate it where it surfaced, treading water as it chewed.

Breakfast was brief, the otter swam below some waterside alders, there was a final hump, the ripples receded, and the reflection of the bare branches reformed.

There are miles of old fence lines lying across the Highlands. Their remains testify to the value in keeping sheep within their bounds, for they must have cost vast sums of money, time and effort to erect. Fortunes were made from sheep-rearing then, and men were hung for sheep-stealing. How many eagles would have been killed for the same reason, guilty or not.

Overleaf:
Lochside reflection of a western moor in autumn.
This colourful plant is purple moor-grass Molinia caerulea, named after its sheen in late summer;
in autumn it is rusty orange coloured, in spring it is a different colour again - bright green. Here
the moor has been heavily grazed and burned, and the grass now dominates it. Before, the ground
cover in such places would have held more heather and herbs, and there would have been pockets
of willow and dwarf birch, creating a richer habitat. There would have been more food for a
variety of animals and in turn more prey for eagles.

Old pines in an old pine wood - in urgent need of regeneration if eagles are to build eyries in their crowns for centuries to come.

The subtle tones of a Highland autumn – the romance of the Highlands.
And eagles are also romanced.

When hunting was at its height in the late nineteenth century, eagles and all other raptors
were not only hunted as vermin, but regarded by some as sport. Trophies did not merely
consist of one adult bird, but the pair, their eggs, chicks and even their eyries were all
desired collectables.

In Highland deer forests, eagles have traditionally been left alone and unpersecuted, as they eat grouse which when calling in alarm can ruin a stalk.

This eagle's eyrie was burned out the year previous to that when the photograph was taken, as can be seen by the scorch marks beneath the nest. Eyries in the adjacent home range have also been burned out, one of them while there were eggs. And I know of other sites where eyries have been burned by fires lit by - whom? shepherds, gamekeepers? Who else would have motive to do so? There are also nests where rocks have dropped onto the eggs – accidentally? And I have heard it alleged that an eagle was kept off its nest by gamekeepers out on the hill nearby with dogs and guns – long enough on a cold day to cause the birds to desert their nest or the eggs to chill and not hatch?

This golden eagle was found dead in its eyrie with two eggs beneath it. It had eaten a lethal dose of alphachloralose, an agricultural pesticide, which subsequently caused it to die of hypothermia while trying to incubate its eggs. The nest was one of a group in a home range with a long history of breeding success. For the estate in which most of the birds' range lay was managed in an amiable way for eagles. Unfortunately, their hunting range would have extended beyond the estate boundaries and the bird could have eaten the poison elsewhere. The poison-laced carcass of a mountain hare was found on a hill opposite the eyrie, on an adjacent estate. Restriction of pesticide use is imposed under the Control of Pesticides Regulations (1986) and the possession of this chemical has been banned under the Possession of Pesticides (Scotland) Order 2005 – together with   several other poisons commonly used to kill birds of prey.

However, since this order, poisoning incidents resulting from deliberate abuse of chemicals have continued and are widespread across Scotland with numerous raptors, other birds, foxes, cats and dogs having been killed. Several golden eagles have been found dead in recent years; in 2010 the abuse of aldicarb was found to be the cause of death of one golden eagle and that of carbofuran for the death of three others and a white-tailed eagle (Science and Advice for Scottish Agriculture).

Other poisons found include sodium cyanide and anyone out in the country should take carewith any dead raptor or other carcasses which might be laced with poison, do not touch them. Report them, or any raptor caught in a trap, immediately to their local police wildlife crime officer who will advise on the best action to take.

I have heard an allegation of one gamekeeper who boasted of killing thirty golden eagles in oneyear. If true, most of those birds would probably have been immature, and as about 130 golden eagle chicks fledge in Scotland per annum, that one   person destroyed the equivalent of more than one fifth of the golden eagles reared in Scotland in one year. What if someone has done this or does so every year? Or if others do likewise? Illegal persecution is the greatest restraint on the eagle population, and there is much vacant ground with plenty food and nest sites, particularly in the eastern Highlands where grouse moors are managed. There are about 440 breeding pairs of golden eagles in Scotland, there could probably be about 500 pairs of breeding golden eagles in Scotland if they were not killed.

Smoke billows from muirburn. This is a very effective method to improve moorland for raising red grouse, as it clears away old growth and encourages fresh shoots of heather - the grouse's main food. This is a perfectly reasonable strategy. However, there is a long history of eagles and other raptors having been relentlessly destroyed on grouse moors where they are regarded as vermin. It is against the law to kill these birds yet persecution of golden eagles seems to continue on some areas managed for shooting of red grouse, and the main gaps in their distribution in Scotland are in such areas. The loss of breeding birds is the main threat to eagles in Scotland. Young birds are likely attracted to vacant home ranges left by any birds killed. Then if the new recruits are killed, the again empty home ranges attract more birds, which are subsequently killed, and so on - to the detriment of recruitment into the wider area of eagle habitat nation-wide.

Is that a reasonable strategy?

Wind farming is a recent industry in the Highlands. Turbines are up and running on numerous sites and more are being planned. In the modern enlightened attitude to land use, proposed sites and running windfarms are assessed and monitored for any effect they might have on golden eagles or other wildlife. Displacement and collision are possible effects on eagles, they have happened in other countries, and both need to be avoided by considerate siting of any wind farms in the Scottish hills.

Ruins of houses, crofts or shielings can be found in most Highland glens. Fewer people may now live there, but there are increasing numbers of visitors – and probably fewer quiet days and places for eagles to hunt and nest in. I know several eyries which are no longer used, probably due to too much disturbance; intentional and unintentional. In the old days, very few people went hillwalking, climbing or bird watching.

22

SNOW STORM

Deep in the hills the winter nights are dark and clear,
but in a whiteout, the way is far from clear.

The stars and the full moon shone bright in the sky and the frost shone bright on the ground. The walk along by the loch that night was almost as clear as by day. Old snow helped to light up the ground here and there and although there was no colour, the form and lie of the land were quite distinct as a few friends and I followed the track's twisting line over the moor.

Once we had turned into the head of the glen there were mountains all around us, the road was far out of sight and mountain ridges framed a wide sky. An hour or so later, as we drew close to our destination, a gable reared clear of the horizon and the bothy grew out of shadow into stone. A pair of rowan trees by the door suddenly appeared like ghosts waving their arms, friendly ghosts, and there was a wink of fire-glow in the window. Some other hillwalkers had arrived before us and set the fire going – such a warm welcome on such a cold night. And it didn't take us long to dump our heavy packs on the floor, say our hellos and snuggle into a spot by the fire. A few more sticks were thrown onto the fire, sending it roaring up the lum behind the old broken fireplace. Snow melted off the toes of our boots, puddles formed on the floor-boards around us, and the walls glowed and flickered with shadows as we

The weather in Scotland includes long periods of rain and snow fall, eagles can sit out such spells for a few days, but if the storms hold up they need to go out and hunt or else starve.

settled down with a warm brew of tea. Outside, silent black clouds softly snuffed out the lights and the black waters of the loch were lost in the dark.

A door creaked open then slammed firmly shut. The fire leapt up and coughed a pillow of smoke out over the room. All I could see from the cobweb grey window was darkness and darkness, but I could hear the rowans' branches bending and rattling in a wind that was not there just a short time before. Then a white fleck shot though the black, then another and soon a whole stream of them were slanting on by. The heat began to seep from the room. And the dark corners grew larger as we grouped tight together in front of the fire. Warmth on our faces. Cold drafts on our backs.

Dawn never truly broke the following day, and a glance through the chilled glass showed nothing except white. All the world was white and becoming deeper in white. It was not a day for the hill so when the showers eased off a little we scrambled up through a broken shelter belt of trees to collect fallen branches for the fire. The lapse didn't last though, and before long we were back feeding the logs onto the flames. Wrapped up in a down sleeping bag, cup in hand, I looked forlornly out of the window. The snow fell all morning then began to break up for a while in the afternoon. I dashed out for a short walk and look around.

During a clear spell I could make out the dark looming forms of crags up on the high ground, and tiny dots of hinds scraping for a nibble of grass on the lower ridges. A peregrine hung on the wind above the hill out the back. It must have been very hungry to be driven out to hunt on such a day. There wasn't much about for it to catch. The only other birds I saw that day were two meadow pipits picking at fat and crumbs thrown out from the door, and a robin and wren flicking through the pile of bent iron in the dilapidated stable next door.

By evening the sky was clearing and the landscape spread out. Slioch, one of the finest of the fine mountains in the north-west, was popping its dome over a distant pass, and the last yellow light backlighting on it was as much of the sun as we were to see. Then the stars came back out. Orion's belt was strapped high, and the moon came up, a big fat full one, fresh and bright after a day's rest. Floodlit white with a velvet black sky behind, Slioch was magnificent. It wasn't so much that it is high, more that its summit was highlit, and peeking over the mid-foreground ridges, which held deep lunar shadows all along their length. Unfortunately we couldn't stand in the clear air to admire the sight for too long. The biting wind once again brought stiff blasts of ice from the north and we retreated for a second stormbound night.

The weather in the morning on the second day was still no better. And it was in response to the frustrations of confinement, rather than any eagerness to go out into what we knew would be heavy-going conditions, that we went onto the hill. Even the first gentle slopes were hard walking as they held deep heather and bog; the moor was full of traps. Deep drifts and holes gulped down any badly

An eagle's view over the moors in winter.

Under such complete snow cover any animals that are not white would be easily spotted by an eagle waiting on high. Even white animals would be seen if they stirred, made obvious trails or cast shadows.

placed foot. Once we had passed the hags, the steep climb up the ridge gave a bit of relief but we were still slow, high stepping through the deep fresh snow.

Gradually, we didn't need to lift up our feet so high and then not at all as we strode quickly over the ice crusted summit. A mountain hare's footprints lay clean-pressed, made only seconds before, although I never saw a glimpse of their printer as spindrift was rushing and hissing along the ridge. The wind rolled snow powder into the tracks and they were gone. Our tracks were gone too. There would be no easy line of retreat to follow. Everything was white on the summit ridge. Not a single dark rock poked through the mantle to take some of the glare off our eyes. Now and again all sense of direction faded as soft cloud blew around us and merged as one with the ground. It was spooky, we could barely see one another when only inches apart, we couldn't even see our feet. And we knew that beyond that soft wall, only a few metres away, out of sight maybe – out of mind definitely not, there was a very long drop down into the corrie below via some very hard rocky ledges.

There was no sharp ridgeline to lead us, so compass and map were braced and sheltered from the wind. We took the quickest of quick bearings and a quick shake of the map to tip out the ice crystals that had drifted into a fold and buried half of the hills in the area. Another short scene of whitescape and whistling wind then followed, before a trace of rime edged boulders showed the way to the top. A sudden strengthening of the wind suggested that we were quite exposed on a rather fine airy perch. Then, right on cue, a hole opened in the mist to reveal a line of high crags hanging over the backside of the hill - only metres away, just as we thought.

A rolling line of cornices snaked all along the top. It was impressive. So impressive that we took a fresh bearing and walked smartly down in the opposite direction. And in what seemed to be no time at all, going downhill, we were back out of the wind and down into a bealach. Then as the visibility increased, we wondered if we had been cheated of a great climb or sensibly cautious. Whichever it was, our descent was expedited by glissading as fast as we could down an open slope, taking care to avoid the crags neglected by the cartographers, but remembered from a previous visit to the area by myself. We rattled and bumped our way down in great roaring fun until the angle eased off and the snow deepened. We ground to a halt, and a solitary snow bunting flushed from a clump of ice-stiff grass heads sticking out of the snow. We were back with the rest of the world.

Down in the glen, the wind was fading and the sheets of cloud and snow were cast off. The sun melted blue holes and they grew larger as the afternoon passed away. We could see right across the glen, and I soon picked out the profile of an eagle as it sat basking in the sun, backlit on the skyline. That bird had probably spent two days and nights cooped up in a rocky niche down in the

A mountain hare basks in winter sunshine for warmth – safely by keeping tucked under a rocky overhang for concealment from golden eagles that might otherwise see it and catch it unawares from above.

icicled overhangs below its now sunny perch. Its feathers would have been fluffed wide and its toes and head drawn in to the down underlayer. Now it would be hungry and it was watching all around, scrutinising every dark shape on the white landscape for prey.

A herd of sixty red deer hinds ploughed criss-cross paths across a corrie, but I never saw feather or hair of grouse or hare. Most, if not all the grouse had moved out from the moor on the approach of the storm. I had watched enormous packs of them flying about the glen as we walked in, and they would be lying up on the lower ground that was still black. All the hares I had seen earlier, and there were plenty of them, were in the shelter of the wood where we had collected the firewood. They would come out and scratch through the snow for food once it was dark.

The storm was over next day and so was our trip. So, after packing up our gear, sweeping out the floor and resetting the fire in the grate for the next users, we closed the door of the bothy and set off down the glen. On our way we passed several small plantations of pine that were old enough to give shelter to the deer. The snow was lying deep there and it was pitted with round yellow bowls. Each hollow had a pile of deer droppings and tufts of hair in the bottom, clearly showing that this was where the deer had sat out the storm. Several of the larger ones had a half-sized bowl next to them. These had been made by calves lying up close to their mothers.

The deer were still in the wood and they scattered off out onto some open hillocks as we approached, their long legs lifting high through the drifts – if only I could do that. Some branches had fallen under the weight of the snow and already the deer, in their hunger, had stripped them of their needles and bark. And the tips of the bottom branches on the trees, which were bent low under the snow, had also been eaten. A party of hoodie crows chattered and jived through the plantation, never stopping in one place for longer than a quick poke or stab through the snow, before rolling on out to the hill in the wake of the deer.

When we were out of the trees we looked up to a pair of eagles right over our heads. One called and curled round in the light breeze to follow our path. The other hung higher up above its mate - waiting on. I suspect that they were watching our progress down the glen, cautiously waiting for us to flush a hare. Both birds tracked us for a about a mile before spiralling around and up to cruise along the hillside above, perhaps to see if the deer there were any better at driving on hares.

The hoodies flew on down the glen ahead of us, seven of them altogether, avoiding the eagles' attentions. They flew low, silent, and straight out of the glen. We walked slowly in a meandering path through the snow.

While walking the hills in thick hill fog I have on several occasions been able to approach within a few metres of golden eagles and heard them call to one another. They have been described as being mute birds, however, after studying them for years I have found that to be wrong. They are really quite vocal, often yelping to one another. Who else has heard their call on a high misty ridge, unaware that they were so close to these spirits of the hills?

23

TIMELESSNESS

When the mist comes down and all that is visible is unaltered by man,
there is a rare sensation of being in a truly wild place.

Thin wispy lines folded into the base of the cliff two hundred metres below. The sheer grey rocks absorbed the waves' beatings one after another as they have done for millions of years. Yet it was those waves and their wind-borne kin, the rains, which have slowly created the mighty Atlantic walls.

There wasn't a sound from the breakers, they were too far away. A thin breeze playing a soft fluting through the cliff-top grasses was the only sound, except for a brief disjointed warble from a trio of skylarks jostling for shelter behind the tussocks. It was only then as I watched the skylarks that I noticed how the first green of the new year's grass was flushing through the old wintry yellow. I sat behind a hummock and watched the strings of fulmars as they tied figures-of-eight and fanciful splices around the foot of the cliff. The fulmars seldom ventured up beyond the main body buttresses. Their world is oceanic and they only come ashore to breed. At the base of the cliff there was a firm band of rock which offered safe nesting ledges. Higher up the face, the rock was rotten with very little security on offer to any animals' limbs. Geologists might have more accurately referred to the rock as weathered, but to a climber it was definitely rotten.

That was where the plants faired better, on the bold ledges, well out of reach of rabbits or even that scourge of mountain plants, goats. Thin moustaches of thrift and scurvy grass filled the window boxes on the sills, and lush beds of wild roses, hazel and willow flourished in the shelter of water-cut gullies. The head of the cliff was held firm and proud, fit to brace the worst that the wild grey westerlies could throw into its face.

Evening drew in and the spring sunshine was soon fading fast. The wind had dropped in the afternoon, and the sea settled down to a languorous rhythm, reflecting the white sun in a long lustrous strip. Distant blue hills on the outer isles climbed out of the ocean's haze to meet the lowering clouds. And slowly, the shine on the Minch's flat water was dulled, as the sun slipped away to the west, to Tir nan Og, the land of everlasting youth.

A gentle fog crept quietly in with the dropping sun. Misty trails drifted past me, and thick pillows spread over the moorland behind. White screens pulled overhead, then faded out, came back, and then slipped away again. It was a soft mist, almost dry, all calm and comforting. There was no impression of height or distance. Everything was veiled and peaceful.

A faint call came from deep in the softness, a meek puppy yelp. I heard it twice more, but could neither fix its position nor put a name to it. It was a call that could have been heard by anyone standing where I stood then; the day before, the year before or hundreds of years before. The softness had hidden the man-made scars on the landscape. Fences, ditches and pylons lay somewhere behind me. Sea-taming lighthouses lay somewhere ahead of me. Mists of the past would have held the same aura as the mists of that day. Old Celtic tales warn of kelpies luring people astray. All with good reason, for if I had followed those plaintive calls I would have been lured over the precipice.

In silence, the kelpie materialised out of the mist. A golden eagle swung less than ten metres overhead, out from the dark edge of the cliff and into the pale yellow halo in the west. Its mate, the slightly smaller male, followed right behind. Magnified by the mist, the duo were stunning, magnificent. Then with a curl of the cloud, both wisped back into ethereal form. There was a quick sequence of yelps like those which I had heard earlier, the eagles were calling to one another in the mist, keeping contact? They have no more magic than us, they are simply better adapted to life in the hills.

I left the eagles to it, dropped downhill out of the cloud and into a long walk across a wet moor. My feet were firm on the ground, they always have been and they still are. As I neared the road in a rising cold wind and drizzle, with water seeping through my jacket and my feet squelching in their own boot-filling puddles, I thought of why I keep going out to these far-flung places in all weather. The reason was simple; it had been a day to remember.

.......It had been yet another eagle day.

And there will be many more..........

A golden eagle slips away into the clouds

One day, while driving across the moors between Torridon and Kishorn, I became quietly aware that an eagle was hunting beside me. It was gliding almost at car window height dipping and tucking back and forth. Then it swung right over the roof and slowly drifted away across the hillside, keeping in rhythm with Beethoven's pastoral symphony which I was listening to at the time - corny but true. Now, whenever I hear that piece of music I see an eagle on the wing.

Further reading

There have been a vast number of words written about eagles, such is their appeal. Below is a selection of books and scientific papers which between them give a broad perspective of eagles and the land they live in. They give details relevant to some of the points mentioned in this book; their status in Scotland, their ecology, and facts relevant to man's use of their land. Even more information can be obtained from the references cited in these publications.

Books:

Forrester, R. & Andrews, I. 2007. Eds. *The Birds of Scotland*, Scottish Ornithologists Club, Aberlady.

Gordon, S. 1955. *THE GOLDEN EAGLE King of Birds*, Collins, London.

Love. J, A. 1983. *The Return of the Sea Eagle*. Cambridge University press, Cambridge.

Newton, I. 1979. *Population Ecology of Raptors*, Poyser, Berkhamsted.

Ratcliffe, D. 1990. *Bird life of mountain and upland*, Cambridge University Press, Cambridge.

Watson, J. 1997. *The Golden Eagle*, Poyser, London. 2nd Edition, 2010.

Watson, J., 2002. Golden eagle *Aquila chrysaetos*. In: Wernham, C.V., Toms, M.P., Marchant, J., Clark, J.A., Siriwardena, G.M., Baillie, S.R. (Eds.), *The Migration Atlas: Movements of Birds of Britain and Ireland*. Poyser, London, UK, pp. 421–422.

Scientific papers:

Campbell, S. and Hartley, G. 2004. Investigation into golden eagle predation of lambs on Benbecula in 2003, *Scottish Agricultural Science Agency*, 2004.

Newton, I and Galbraith, E.A. 1991. Organochlorines and mercury in the eggs of Golden Eagles *Aquila chrysaetos* from Scotland. *Ibis*, 133: 115-120.

Watson, A., Payne, S. & Rae, R. 1988. Golden Eagles *Aquila chrysaetos*: land use and food in northeast Scotland. *Ibis*, 131: 336-348.

Watson, J., Rae, S.R. & Stillman, R. 1992. Nesting density and breeding success of golden eagles *Aquila chrysaetos* in relation to food supply in Scotland. *Journal of Animal Ecology*, 61: 543–550.

Watson, J., Leitch, A.F. & Rae, S.R. 1993. The diet of golden eagles *Aquila chrysaetos* in Scotland. *Ibis*, 135: 387-393.

Whitfield, D.P., MacLeod, D.R.A., Watson, J., Fielding, A.H. & Haworth, P.F. 2003. The association of grouse moor in Scotland with the illegal use of poisons to control predators. *Biological Conservation*, 114: 157–163.

Websites

The following websites give further information and news on eagles in Scotland:

Scottish Raptor Study Groups,
http://www.scottishraptorgroups.org

Satellite tracking,
http://www.natural-research.org/latest-news-research-updates/satellite-tracking-in-2010/

Poisoning incidents, *Science and Advice for Scottish Agriculture,*
http://www.sasa.gov.uk/animal-poisoning-reports

Partnership Against Wildlife Crime Scotland (PAWS),
http://www.scotland.gov.uk/Topics/Environment/Wildlife-Habitats/paw-scotland

Royal Society for the Protection of Birds,
http://www.rspb.org.uk including
http://www.rspb.org.uk/Images/Bird_of_prey_persecution_tcm9-289614.pdf

Scottish Natural Heritage,
http://www.snh.gov.uk

While filming two cock Capercaillie displaying at a lek in Norway in 1999, Odd Aukland captured footage of a golden eagle flying in and attacking one of the cocks, eventually killing it, although the bird fought back vigorously. Then while it mantled the dead bird the second cock attacked the eagle, which promptly turned and killed it too. The second bird had probably diverted its aggression which had been focused on its sparring partner, towards the eagle, driven by high levels of testosterone running through its body during the lek display. The eagle in the film looks like a male by size, and is clearly adept at killing such large prey.
The film can be viewed on line at:
http://www.youtube.com/watch?v=nb1H_-S4Xjk

Information and news on the author can be followed at:-

Stuart Rae, Wildlife and Wild Places,
http://sites.stuartrae.com
http://stuartrae.blogspot.com

The best way to see eagles is to watch the sky.
I might not always see eagles but I've seen some fantastic skies.

For your notes and observations:

Forthcoming Titles in the Birds and People (BP) Series.

1 Eagle Days - Stuart Rae

2 Red Kites - Ian Carter

3 Short-eared Owl - Don Scott

4 On the Rocks (sea birds) - Bryan Nelson and John Busby

5 Harriers and Honey Buzzards - Mike Henry

6 Derek Ratcliffe - Various authors

Others to follow in the near future.